FROM BIG OIL TO BIG GREEN

FROM BIG OIL TO BIG GREEN

Holding the Oil Industry to Account for the Climate Crisis

MARCO GRASSO

The MIT Press
Cambridge, Massachusetts
London, England

The MIT Press would like to thank the anonymous peer reviewers who provided comments on drafts of this book. The generous work of academic experts is essential for establishing the authority and quality of our publications. We acknowledge with gratitude the contributions of these otherwise uncredited readers.

This book was set in Scala Pro and Scala Sans Pro by Westchester Publishing Services. Printed and bound in the United States of America.

Library of Congress Cataloging-in-Publication Data

Names: Grasso, Marco, author.
Title: From big oil to big green : holding the oil industry to account for
    the climate crisis / Marco Grasso.
Description: Cambridge, Massachusetts : The MIT Press, [2022] | Includes
    bibliographical references and index.
Identifiers: LCCN 2021035152 | ISBN 9780262543743 (paperback)
Subjects: LCSH: Climatic changes. | Oil industries.
Classification: LCC QC902.8 .G73 2022 | DDC 363.738/26–dc23/eng/20211014
LC record available at https://lccn.loc.gov/2021035152

10 9 8 7 6 5 4 3 2 1

Let me give you a definition of ethics: it is good to maintain and further life, it is bad to damage and destroy life.

—Albert Schweitzer

Power concedes nothing without a demand. It never did and it never will.

—Frederick Douglass

Written laws are like spiders' webs; they will catch, it is true, the weak and the poor, but would be torn in pieces by the rich and powerful.

—Anacharsis

CONTENTS

CONTENTS

ACKNOWLEDGMENTS

I thank Sabina Zambon for the countless unvaluable insights into the arguments of the book; her painstaking editorial, linguistic, and fact-checking work; her efforts to untangle my academic jargon; and her occasional additions that provide overall color. Without her, this book would have been very different and indeed much less interesting and enjoyable.

I thank the Polish Institute of Advanced Studies (PIASt) of the Polish Academy of Sciences for having provided me with the means—including a cozy apartment along the Vistula—and the much-appreciated peace and quiet to finish this book over the course of a difficult locked-down period.

I thank Richard Heede for having helped me collate the Big Oil figures.

I thank Beth Clevenger and the MIT Press team for their continuous support throughout the project.

INTRODUCTION

The word *crisis* has its roots in ancient Greek, *krisis*, meaning the tipping point in a predicament, the moment when paths diverge, only one leading to recovery. This is a concept that the COVID-19 pandemic, which exploded in early 2020, has made us chillingly familiar with. A world in which carbon emissions are either brought under control or allowed to flourish unabated is also at just such a path, and science, the media, and politics—not to mention social networks—abound in hellish portrayals of possible future scenarios.

And like the virus that threw the world into a crisis at the beginning of the 2020s, emissions possess no passports and respect no political or natural borders. Both COVID-19 and carbon emissions bring rise not only to a senseless loss of lives but also to colossal expenditure in economic terms.

This book will not dwell on the pandemic, limiting its reflection on it, in the concluding chapter, to the implications the COVID-19 crisis has had and will have on the oil industry; however, a consideration of the shared patterns of inequalities deeply embedded in our society is conducive to underlining the moral perspective behind this book's premise.

A moral perspective is a fundamental requirement for ensuring a broad spectrum of backing for enduring strategies, policies, and norms to be implemented in collaboration with the main polluter so as to achieve the kind of effective long-termism required that differentiates a marathon from a sprint.

☆ ☆ ☆

Oil permeates our lives in endless ways. It is everywhere, in clothing, in furniture, in computers and smartphones, in those minuscule granules in toothpaste that give us that extra-fresh feel, in the gloss we smear on our lips, in the medicines that cure our ills; it is the global economy's primary energy and fuel source. Oil also lubricates the global supply chains that bring us Earth's bounty. Even one of the simplest consumer products imaginable, a plain white cotton T-shirt—a mainstay of everyone's wardrobe, from hipster Brooklynites to the humble immigrant tomato pickers in southern Europe, seemingly oil-free—is a masterful triumph of global cooperation and coordination. And it is brought to us by none other than Big Oil itself. Cotton is planted, cultivated, and harvested in Mississippi with oil-based chemicals and machinery; then it is sent through oil-powered shipping vectors to spinning factories in Indonesia; the spooled yarn travels on oil-propelled vessels to garment factories in South Asia and Latin America. Finally, the global shipping industry that is the very foundation of the entire global consumer economy—it takes only fifteen supervessels powered by dirty high-sulfur heavy fuel oil to emit as much sulfur as all the world's cars and as much greenhouse gas as 760 million cars–brings the unassuming cotton T-shirt to a store near you.

Greenhouse gases, mainly in the form of carbon dioxide, have been on the rise since the industrial revolution started, almost 250 years ago; the atmospheric concentration of these gases is higher than it has been in the past 800,000 years, with billions of tonnes of carbon dioxide being released annually from the combustion of fossil fuels. Along with the felling of carbon-absorbing forests for timber, cattle grazing, and crop harvesting, this is one of the main causes of the changes in the climate that are disrupting our world. The temperature on Earth has been climbing steadily for the past two centuries, but since the post–World War II boom in both consumption and population, it has rocketed. This increase in energy in the atmosphere is provoking a surge in extreme weather events in terms of both frequency and intensity. Trillions of tonnes of ice have melted, and in mountain ranges the world over glaciers are shrinking, causing inexorable sea-level rises with the consequent drought, floods, and heat waves, casting millions into despair. These are just a few of the most palpable ramifications on Earth itself without even taking into consideration the effects on wildlife. Biologist Mark Urban (2015) estimates that approximately one-sixth of species face extinction due to climate change.

To avert the most disastrous impacts of these extreme changes in weather patterns, the 2015 Paris Agreement (within the United Nations Framework Convention on Climate Change) set an aspirational target to pursue efforts to limit global temperatures to 1.5°C above preindustrial levels. How close we can get to averting this level—and avoid plunging Earth into possibly irreversible catastrophe—depends very much on whether or not we can create systems that diverge from our current fossil fuel–reliant path. Fossil fuels need to be phased out and replaced by zero-carbon alternatives, as current socioeconomic systems are likely to tip Earth past that 1.5°C threshold.

It is not the physical impossibility that is the obstacle to achieving this goal; it is the choices made by global society, one in which governments and industry continue to oil the machine of this threat to our climate and our health. There is an emerging focus on how tackling the supply chain of fossil fuels can impact climate change, one that previously took a back seat to the usual preoccupation of the significance of demand and consumer-based emissions; indeed, more enlightened politicians are starting to go beyond mere rhetoric, claiming that they will create proposals to limit harmful fossil fuels, with the idea of *keeping them in the ground* being openly advanced by some of them for the very first time.

Oil executives too are changing their tone on climate, making tentative steps toward recognizing that their products have a negative effect on the environment, and yet they continue to plan a future in which fossil fuels play a major role. They proclaim a desire to be part of the solution, all the while being at the helm of the problem. But the companies themselves, despite their countless declarations and pledges, seem to have no real intention to look their gift horse in the mouth for the foreseeable future. The International Energy Agency (IEA) reports that in 2019 the oil industry's capital investments in fossil fuels were 99.2 percent of total energy investments, compared to a mere 0.8 percent of those in renewables and carbon capture and storage (IEA 2020b). Evidence shows that the world's fifty biggest oil companies are going to flood the planet with an additional seven million barrels of crude oil per day over the coming decade. A report published by the Global Gas and Oil Network stated that in the four-year period between 2020 and 2024, the industry as a whole plans to invest $1.4 trillion[1] in extraction projects alone. To consider just one example of an individual company, according to September 2021 internal estimates, ExxonMobil is expected to produce more than 9 billion barrels of oil equivalent offshore of Guyana.

No one could deny that Big Oil's role in our lives is astronomic as also testified by the constant tsunami of news reports regarding the oil world, most of which provide an at most superficial glimpse into the industry's tortuous initiatives aimed at protecting their interests. Keeping abreast of the bigger picture is a painstaking task.

An aggravating factor for those most incontrovertibly involved in pumping huge amounts of carbon into the atmosphere is denialism, one of the tenets upon which Big Oil's moral responsibility for the climate crisis rests, as this book asserts. While most oil companies now acknowledge that fossil fuel combustion is causing climate change, many continue to further their financial interests by funding disinformation about the role they play in it, how much of it is even caused by anthropogenic actions, and the extent of its harm to human health. Major industry players pour millions into groups such as Energy In Depth, a self-proclaimed "research, education and public outreach campaign" that tries to undermine science and discredit scientists critical of the oil business, waging outright war against new environmental regulations all while "straightening out the myths you may have heard about what we do and how we do it [and] how the shale *revolution* in the United States continues to impact energy markets (for the better!) all around the world" (Energy In Depth 2021).

This kind of convivial, benevolent language is typical of the virtue signaling by oil companies in their efforts to demonstrate their environmental rectitude and is part of the all too common parade of greenwashing whereby the public is deceived about the environmental benefits of a particular product, service, or habits of industries. The same effect can be had by promising to offset emissions, a public laundering of the conscience and of reputations that contributes to undermining policies to tackle the root cause of the harm, by burying the problem under a public-friendly Band-Aid. So while Big Oil executives promise to take (baby) steps away from a century and a half of despoilment, they have still—to draw an illustrative parallel—run up a huge tab on their credit card. And although pledging to make fewer purchases, they remain bewildered that the balance owed does not disappear, does not miraculously turn into water under the bridge, all forgiven. Because in the future, the interest will continue to accrue on those emissions, both past and present.

One of the industry's most masterful strokes was to persuade the public that anthropogenic climate change is a question of individual consumer choice, that the word *energy* is synonymous with *fossil fuels*, that a world without their particular brand of energy would catapult humanity back to the Stone Age. Fossil fuels are basically just meeting a demand, they proclaim. This illusion of consumer choice was produced by a well-orchestrated sleight of hand during the postwar boom years. Ingenious promotional campaigns employing the best advertising minds—as well as swaths of a complicit media made up of journalists whose main ability was to ventriloquize rather than question—meant that the system created in the industry's own image became embedded in the pervasive social, political, and economic infrastructure that, in truth, made environmentally friendly choices all but impossible. This narrative has blurred the reality of the climate change debate for the past few decades, laying the blame squarely on the shoulders of a public who loses sleep over forgetting to bring reusable totes to the supermarket, not replacing their halogen light bulbs sooner, or driving the kids to school rather than having them walk, casting consumers as saints or sinners depending on the extent to which they can adhere to these ethics. This is not meant to deny the power and wherewithal of the consumer or forget that both governments and industry interpret and react to consumer demands, just that it is harder to provoke a sea change when habits are so deeply ingrained or when the concept of alternatives is all but nonexistent in our imaginations.

The COVID-19 pandemic has underlined how the globalization of policy and innovation can evolve—and how fast—to ensure preparedness for inexorable future widespread emergencies. We have now witnessed firsthand how the political compass can be shifted rapidly when necessary and how willing the public is to get behind policy making that, while initially provoking opposition, is embraced as being for the greater good once comprehension is established.

But this is not about altruism—the world is starting to wake up to the fact that our actions can come back to deal us a blow in the face with all the force of a right hook from a heavyweight boxer—and instead is about self-interest. It is about using the moral argument persuasively to achieve the means to an end, with the *means* being a substantial financial injection to right Big Oil's wrongs that have led to the climate crisis (reparation) and the greening of its business (decarbonization) and the *end* being an averted catastrophe.

The climate crisis calls for a coordinated, collaborative global effort. Leaders must understand that dealing with it successfully requires measures that are not just tip-of-the-iceberg and concern the immediate future but are also systematic and far-reaching in time. There can be no *either/or* about this issue; it requires a resolute *and also*. It concerns everyone, everywhere, and from this moment forward. Governments and the industries they support can no longer cherry-pick the facets of globalization that best suit those for whom the cherries are within easy reach.

* * *

In London in 1961, the British philosopher Bertrand Russell, almost ninety years old at the time, led a march calling for nuclear disarmament. His gaunt frame topped by a shock of snow-white hair, he paused to rest on the steps of the UK Ministry of Defence on Whitehall. A BBC news reporter asked him why, at this stage in his life, he was exerting himself with protests. In his impossibly posh voice, he retorted, "Well, if the policies of the present government are continued, they will inevitably result in the destruction of the human race. And some of us think that is rather a pity" (Morgan 2020).

Russell's thoughts could be echoed by movements gaining traction such as Extinction Rebellion and Greta Thunberg's Fridays for Future, which have seen protests spilling out of nongovernmental and international summits onto the streets and social media, each and every day, with the goal of causing ruptures within the oil world. This book christens such movements, political authorities, economic and financial entities, and charismatic individuals as *agents of destabilization*. They are the fruits of a great generational endeavor determined to hold polluters accountable and are fueled by an impatient millenarian quality, of an expectation, a demand, to witness a momentous sublimation of harmful practices.

Agents of destabilization's charges against the industry regarding climate change are not only raising public awareness but are also starting to cause ripples of concern among financial investors. *Business as usual* is showing its cracks, starting to appear obsolete, moribund. The kind of snail's pace, incremental progress the oil companies tout could prove too little, too late. Campaigns for divesting from fossil fuels are proliferating worldwide and being adhered to by a range of big investors, such as sovereign wealth funds, pension funds, banks, universities, media platforms, and religious institutions. Similarly, initiatives to coerce fossil fuel companies to keep their reserves underground are multiplying. The Keep It In The Ground movement even seems to have garnered the approval of Pope Francis, investigative journalism is shedding light on unethical practices within the oil world, and oil and gas companies are increasingly being targeted by climate resolutions and lawsuits.

These agents of destabilization are promoting innovative and headline-grabbing democratic and nonviolent road maps to force the industry to consider climate change as the existential threat that it is rather than an unfortunate by-product of its core business, mere collateral damage. Big Oil, by indiscriminately charting the course of the global economy along fossil fuel–reliant paths, is the driving force behind the current carbon-intensive socioeconomic system, and yet it has somehow managed to dodge a significant bullet, so to speak. While much opprobrium aimed its way revolves around the demand that the industry decarbonize, Big Oil has adroitly avoided being cornered into funding action to redress past harm.

In joining a growing chorus of voices calling for the oil industry to decarbonize, this book adds to it a requirement for the oil industry to make amends for said past harm. Raise the topic of reparations in any field—be it slavery, apartheid, or climate change—and a barrage of questions inevitably follows: Who will pay? Who will receive financial rectifications? How will the money be spent? But if the practicalities of reparations are a sticking point, a departure point to reaching a solution has to be identifying the justice behind it, one of the key issues this book tackles.

To clarify, this book does not aim to downplay the responsibility or importance of other agents such as states, consumers, civil society, businesses, and stakeholders, all of whom play a vital role in defining the much-needed initiatives to support climate efforts. However, a novel, more effective—not to mention financially compelling—approach to dealing with the disastrous consequences of the climate crisis can be had in calling attention to the significant role oil and gas companies have played in engendering the climate problem and placing a burden on them to urgently help make amends. This book does not claim to be the first to have pointed the finger at Big Oil for its role in the climate crisis, nor is it the first time someone has claimed that oil companies should pay for the harm actually caused by their operations. But this book's main original contribution to the debate is that it is perhaps the first time there has been an attempt to say "Right, this is the moral framework of responsibility, this is what the industry should do to financially rectify the harm it has caused, and (not insignificantly) this is how much cash they should stump up to do so."

Yes, the frequently encountered stumbling block of the classic 5Ws (*who, what, when, where, why*) required to resolve any issue are dissected in great depth in the book, with an additional *how* added into the mix for good measure. With regard to reparations, for example, the book will explain how to fairly apportion responsibility while meeting the needs of duty recipients. The creation of a fund is suggested based on binding international agreements and initiatives, capitalized with disgorged funds from the oil industry. The cash injection from the wealthy agents is channeled into three directions to cover the rectification of harm endured by the most socially vulnerable to climate change on a global scale, to support the low-carbon transition, and to ensure that workers and communities currently supported by the oil industry are not ignored or left behind.

Of course, the different social situations (to varying degrees favorable or less so to the oil business) the oil majors operate in need to be taken into account, as chapter 7 does, separating the top twenty companies into three distinct groups with different ensuing requirements in terms of the duties of reparation and decarbonization based on their respective social, institutional, economic, political, and operational contextual circumstances and via an algorithm that takes into account their assets, historical greenhouse gas emissions, and their responsibility in accordance with the morally relevant facts examined in the first part of the book. The duties are, for instance, less stringent for companies whose revenues from fossil fuel products are used to the benefit of their societies, as in the case of national oil companies owned or participated in by less wealthy countries, and more stringent for international oil companies that, as with American oil companies, greatly contributed to denialism.

In sum, policy implications of the duties of reparation and decarbonization examined in this book illustrate how such duties are first examined though the lens of societal, economic, and political context, with the addition of considerations of how prepared the country in which these companies operate is to break free of high-carbon models of growth, again with a more lenient approach for those companies that play a significant social function in the development of their host countries. These objective considerations will be collated and scrutinized so as to formulate *personalized* reparation and decarbonization objectives and timelines for the major global oil companies while justifying this book's thesis that a managed decline of Big Oil's involvement in fossil fuel production and distribution can best be achieved through efforts to change social/moral norms and raise awareness. And perhaps if Big Oil toes the line by willingly putting its own house in order, it could see itself maintaining a social license to operate that could see it transform into a *Big Green* energy provider.

✻ ✻ ✻

These words from a speech given by Frederick Douglass on the twentieth anniversary of the emancipation of slaves in the West Indies ring loud today with regard to the power of disruptive action: "The whole history of the progress of human liberty shows that all concessions yet made to her august claims have been born of earnest struggle. . . . This struggle may be a moral one, or it may be a physical one, and it may be both moral and physical, but it must be a struggle. Power concedes nothing without a demand. It never did and it never will" (Douglass 1857).

1 THE CLIMATE CRISIS: ALL ROADS LEAD TO BIG OIL

Governments, the media, and the public at large have started to wake up to the fact that the term *climate change* does not quite cover the gravity of the issues at stake, with the more urgent term *climate crisis* creeping into our everyday lexicon.

On January 24, 2017, US president Donald Trump signed a presidential memorandum finally bringing the Keystone XL and Dakota Access oil pipelines to completion.[1] The $3.7 billion Dakota Access oil pipeline stretches 1,172 miles, weaving its way under the American Midwest and thrumming to the sound of the 470,000 barrels of crude oil it moves each and every day: it begins in the Bakken shale oil fields in northwest North Dakota and continues through South Dakota and Iowa to an oil tank farm in southern Illinois. The no- longer-active $8 billion Keystone XL Pipeline was the fourth phase of the Keystone Pipeline System and upon completion would have stretched nearly 1,200 miles from its source at the oil sands in Alberta, Canada, crossing the US border to straddle Montana, South Dakota, Nebraska, Kansas, Oklahoma, and Texas and carrying more than 800,000 barrels per day of carbon-heavy petroleum to the destination point at the refineries on the US Gulf Coast.

These two pipelines were (and the Dakota pipeline still is) at the center of fierce protests by multiple actors across the international stage, as their deployment would wreak havoc on the environment and threaten cultural heritages as well as perpetuate the production and economic reliance on fossil fuels.

On the other side of the Atlantic in prosperous Norway, a country not usually associated with civil unrest, activists literally took to their kayaks on July 21, 2017, to encircle an offshore rig contracted by Statoil (renamed Equinor in 2018) in a remote Arctic area in the Barents Sea, where the company was drilling for oil and gas deposits. On September 27, 2017, Brazilian campaigners gathered in front of a hotel in Rio de Janeiro, where the National Agency of Petroleum, Natural Gas and Biofuels was auctioning off concessions for new oil and gas investments. Similar protests against the industry have been gathering force all around the globe in recent years, from New Zealand to the Philippines to Nigeria to the United Kingdom and Italy.

Is some kind of gearshift under way? Why is it that almost from one day to the next, an industry that, despite its enduring central role in supporting the economy that provides our comfortable lifestyles, has been recast as the global villain? Of course, a hesitant accusatory finger has long been aimed Big Oil's way over broader issues such as environmental degradation, economic exploitation, social disruption, political instability, and cultural estrangement. However, increasingly frequent and pressing is the concern regarding the climate, so the answer—not so surprising—largely lies with Big Oil's contribution to the climate crisis.

Global emissions rose globally by around 2 percent in 2017, despite an encouraging slowdown of the previous three years (Le Quéré et al. 2018), and a further 1.9 percent in 2018 compared to 2017 (Crippa et al. 2019). In 2019 emissions were 4 percent higher than in 2016, when the Paris Agreement was eventually signed (Jackson et al. 2019). Moreover, the 2019 World Energy Outlook of the International Energy Agency (IEA) observed that emissions are on track for further increases until 2040 unless governments take radical action (IEA 2019c), while its 2020 report (IEA 2020e) clarifies that without a systemic change in government policies, oil demand is unlikely to decline and indeed might rise substantially in the next few years; its 2021 global demand forecast is 96.7 million barrels per day, 5.7 million barrels per day above 2020. The 7 percent decrease in global fossil CO_2 emissions compared to 2019 levels caused by the COVID-19 pandemic (Le Quéré et al. 2021) will have limited impact on long-term climate goals and may be followed by a swift rebound (Le Quéré, Jackson, and Jones 2020a) unless countries take rapid action to limit fossil fuels.

But while no one would dispute the fact that fossil fuels remain the driving force behind the world's socioeconomic systems, an increasing number of people are waking up to the prominent role Big Oil plays in the climate crisis, as this book shows.

Yet in proportion to its power, involvement, wealth, and possibilities and despite its star billing in a system subsidizing and promoting fossil fuels, Big Oil is proving itself to be a masterful chameleon in the current global climate debate, assuming a greenish hue when waving the environmental banner before returning to a sparkling gold when it comes to money matters. Nonetheless, possibly in light of Big Oil's role in wealth creation, governments seem unready or unwilling to bite the hand that feeds them.

One reason for this may be the seeming helplessness of tracing carbon emissions to specific companies. However, this lack of empirical data is gradually being overcome. Richard Heede, director of the Climate Accountability Institute's Carbon Majors project, along with a number of scientists, published findings providing overwhelming grounds for an investigation into Big Oil's contribution to climate change in terms of emissions, thus paving the way for an analysis about its responsibility and its consequent duties.

It is worth noting at the outset that this book chooses to distinguish the concepts of *responsibility* and *duty*. The first is the *condition* of being responsible according to principles of justice and the obligation to take action. A duty is a standard of moral behavior imposed by responsibility and involves a *practical commitment* to either undertake or refrain from undertaking specific courses of action.

A further disambiguation is required: when the book refers to *Big Oil*, the *oil and gas industry* (or sometimes only the *oil industry*), or the *fossil fuel industry*—despite the many terminological disputes within the oil world—the reference is to the large multinational companies that engage in the exploration, production, refinement, and distribution of hydrocarbons (i.e., conventional oil, unconventional oil, and unconventional liquids). Conventional oil is the most easily accessible family of hydrocarbons and accounts for the greatest share of global liquid fuels and will likely still account for around 90 percent in 2030. Conventional oil basically includes crude oil, condensate, and natural gas liquids. Unconventional oil refers to less accessible resources that require, as the name suggests, *unconventional* techniques and includes extraheavy oil, oil shale, oil sands, and tight oil. Finally, unconventional liquids are those liquid hydrocarbons produced synthetically, such as coal-to-liquids, gas-to-liquids, and biofuels. It should be noted also that the term *Big Oil* is often used in a derogatory way by the media and detractors to underline the enormous economic and political clout these companies wield, not least due to their lobbying influence, and the ironclad grip their products hold on industrial society. While in the majority of the book the term is used in its broadest possible sense, part III narrows the definition to refer solely to the top twenty public and privately owned oil behemoths in terms of contributions to greenhouse gas emissions between 1988 and 2015.

Oil and gas have given rise to a megalithic business: they are explored, produced, refined, and distributed throughout the globe by a host of industrial titans. These companies, through the emissions generated by their products and processes, have significantly added to the increase in the concentration of greenhouse gases, especially carbon dioxide and methane, in the atmosphere. The relationship between emissions, concentrations, and climate change is well established in the pertinent scientific literature, and science almost unanimously affirms that climate change is directly and profoundly harming the planet and humanity (IPCC 2021a). Therefore, it is possible to surmise that the oil and gas industry has been a key direct contributor to anthropogenic climate change and the domino effect of harm we are witnessing on Earth.

Society and the policy community are beginning to sit up and take notice, with a rapidly growing global concern provoking broad societal and political pressure, shining the spotlight on the entire industry and eating away at its influence at least on a superficial level. Campaigns for divesting from fossil fuels, for instance, are proliferating worldwide: as of September 2021, 1,333 institutions were diverting $14.58 trillion from the fossil fuel industry, whereas more than 58,000 individuals had decided to divest $5.2 billion. Similarly, initiatives to coerce fossil fuel companies to keep their reserves underground are multiplying; even Pope Francis, at a meeting with the energy majors at the Vatican on July 9, 2018, urged the industry leaders to *keep it in the ground*. Investigative journalism is shedding light on the least accessible corners of the oil world,[2] and oil and gas companies are increasingly being targeted by shareholder climate resolutions and lawsuits. A groundswell of opposition to the industry is gathering force, both in the public sphere and from influential voices in a position to shift opinions.

Before uprooting the whys and the wherefores of the oil companies' direct contribution to greenhouse gas emissions, this chapter provides an overview of the oil and gas industry and then analyzes how and why the industry acknowledged climate change.

THE STRUCTURE OF THE OIL AND GAS INDUSTRY

The late 1980s saw the structure of the oil industry—basically made up of privately owned international oil companies (IOCs) and state-owned national oil companies (NOCs, see table 1.1)—morph into its current form, even if over the years the relevance and power of NOCs have significantly increased (see tables 1.2, 1.3, 1.4, 1.5). NOCs control roughly 90 percent of the world's oil and gas reserves (considering only the top

Table 1.1

Largest national oil companies' ownership ($ billion)

NOC	Ownership	Total Assets
CNPC/PetroChina	China (100%)	608.1
Saudi Aramco	Saudi Arabia (100%)	398.3
Gazprom (Russia)	Russia (50.23%)	352.7
Sinopec (China)	China (100%)	317.6
Petrobras (Brazil)	Brazil (64%)	229.7
PDVSA (Venezuela)	Venezuela (100%)	226.8
National Iranian Oil	Iran (100%)	200.0
Abu Dhabi National Oil—ADNOC[a]	UAE (100%)	153.7
Kuwait Petroleum	Kuwait (60%)	136.5
Pemex (Mexico)	Mexico (100%)	101.8

[a] Estimate provided by Richard Heede, dir., Climate Accountability Institute, email communication, April 15, 2020.

Source: Companies' websites; *Oil & Gas Journal* (2020a, 2020b); *Fortune* (2020).

Table 1.2

Oil and gas companies by revenues, 2019 ($ billion)

Oil and Gas Company	Revenues	Typology
Shell (UK/Netherlands)	404.3	IOC
CNPC/PetroChina	364.2	NOC
Saudi Aramco	329.8	NOC
BP (UK)	282.6	IOC
ExxonMobil (USA)	264.9	IOC
TotalEnergies (France)	176.2	IOC
Chevron (USA)	146.5	IOC
Gazprom (Russia)	118.4	NOC
ENI (Italy)	78.5	IOC
Petrobras (Brazil)	76.6	NOC

Source: *Oil & Gas Journal* (2020a, 2020b).

Table 1.3

Oil and gas companies annual production, 2019

	Oil[a]	Gas[b]	Gas[c]	Oil & Gas[c]	
Oil and Gas Company	**Mb**	**Bcf**	**Mboe**	**Mboe**	**Typology**
Saudi Aramco	4.096	3,277	546	4,642	NOC
Gazprom (Russia)	491	17,867	2,978	3,469	NOC
National Iranian Oil	860	8,962	1,494	2,354	NOC
Rosneft (Russia)	1,674	2,366	394	2,068	NOC
Abu Dhabi National Oil[d]	1,278	2,510	418	1,696	NOC
PetroChina	909	3,908	651	1,560	NOC
Shell (UK/Netherlands)	658	4,230	705	1,363	IOC
BP (UK)	807	3,322	554	1,361	IOC
ExxonMobil (USA)	740	2,434	406	1,146	IOC
Kuwait Petroleum	977	493	82	1,059	NOC
TotalEnergies (France)	610	2,688	448	1,058	IOC
NNNC—Nigeria	735	1,689	282	1,017	NOC
Chevron (USA)	550	2,357	393	943	IOC
Sonatrach (Algeria)	373	3,164	527	900	NOC
Petrobras (Brazil)	755	834	139	894	NOC
Lukoil (Russia)	662	1,196	199	861	NOC
Pemex (Mexico)	688	870	145	833	NOC
Equinor (Norway)	363	2,037	340	703	NOC
ENI (Italy)	326	1,930	322	648	IOC
PDVSA (Venezuela)	370	726	121	491	NOC
Total	**17,922**	**66,860**	**11,143**	**29,065**	

[a] Oil: Million barrels (Mb)

[b] Gas: Billion cubic feet (Bcf)

[c] Gas, oil and gas: Million barrels oil equivalent (Mboe, 1 barrel = 6,000 cubic feet).

[d] Estimate provided by Richard Heede, dir., Climate Accountability Institute, email communication, April 15, 2020.

Source: Oil & Gas Journal (2020a, 2020b).

Table 1.4

Oil and gas companies reserves, 2019

Oil and Gas Company	Oil[a] Mb	Gas[b] Bcf	Gas[c] Mboe	Oil & Gas[c] Mboe	Typology
National Iranian Oil	208,600	1,200,252	200,042	408,642	NOC
PDVSA (Venezuela)	303,806	200,372	33,395	337,201	NOC
Saudi Aramco	227,630	190,575	31,763	259,393	NOC
Abu Dhabi National Oil	92,200	200,000	33,333	125,533	NOC
Gazprom (Russia)	10,452	625,591	104,265	114,717	NOC
Kuwait Petroleum	101,500	63,000	10,500	112,000	NOC
NNNC—Nigeria	36,890	203,449	33,908	70,798	NOC
Sonatrach (Algeria)	12,200	159,054	26,509	38,709	NOC
PetroChina	7,253	76,236	12,706	19,959	NOC
BP (UK)	11,478	45,601	7,600	19,078	IOC
ExxonMobil (USA)	13,108	32,924	5,487	18,595	IOC
Rosneft (Russia)	3,935	74,380	12,397	16,332	NOC
Lukoil (Russia)	12,015	21,773	3,629	15,644	NOC
TotalEnergies (France)	6,006	36,015	6,003	12,009	IOC
Shell (UK/Netherlands)	5,264	33,821	5,637	10,901	IOC
Petrobras (Brazil)	8,092	8,549	1,425	9,517	NOC
Chevron (USA)	4,771	26,587	4,431	9,202	IOC
Pemex (Mexico)	5,961	6,352	1,059	7,020	NOC
ENI (Italy)	3,601	19,832	3,305	6,906	IOC
Equinor (Norway)	2,575	17,355	2,893	5,468	NOC
Total	**1,077,337**	**3,241,718**	**540,286**	**1,617,623**	

[a] Oil: Million barrels (Mb)

[b] Gas: Billion cubic feet (Bcf)

[c] Gas, oil and gas: Million barrels oil equivalent (Mboe, 1 barrel = 6,000 cubic feet).

Source: Oil & Gas Journal (2020a, 2020b).

twenty companies in 2019 more than 95 percent; see table 1.4), while in 1970 they had direct access to a limited portion of such reserves—75 percent of global oil production (considering only the top twenty companies in terms of production in 2019 almost 78 percent; see table 1.3)—and own great swaths of the infrastructures (Victor, Hults, and Thurber 2012a; Bridge and Le Billon 2017). However, the reading of these figures is a somewhat thorny issue: NOCs often control reserves that are produced

Table 1.5

Reserve to production ratio,[a] 2019 (years)

Oil and Gas Company	Oil	Gas	Oil & Gas	Typology
PDVSA (Venezuela)	821.1	276.0	686.8	NOC
National Iranian Oil	242.6	133.9	173.6	NOC
Kuwait Petroleum	103.9	127.8	105.7	NOC
Abu Dhabi National Oil—ADNOC	72.1	79.7	74.0	NOC
NNNC (Nigeria)	50.2	120.5	69.6	NOC
Saudi Aramco	55.6	58.2	55.9	NOC
Sonatrach (Algeria)	32.7	50.3	43.0	NOC
Gazprom (Russia)	21.3	35.0	33.1	NOC
Lukoil (Russia)	18.1	18.2	18.2	NOC
ExxonMobil (USA)	17.7	13.5	16.2	IOC
BP (UK)	14.2	13.7	14.0	IOC
CNPC/PetroChina	8.0	19.5	12.8	NOC
TotalEnergies (France)	9.8	13.4	11.4	IOC
ENI (Italy)	11.0	10.3	10.7	IOC
Petrobras (Brazil)	10.7	10.3	10.6	NOC
Chevron (USA)	8.7	11.3	9.8	IOC
Pemex (Mexico)	8.7	7.3	8.4	NOC
Shell (UK/Netherlands)	8.0	8.0	8.0	IOC
Rosneft (Russia)	2.4	31.4	7.9	NOC
Equinor (Norway)	7.1	8.5	7.8	NOC

[a] The reserves to production ratio (RPR or R/P) is the remaining amount of oil and gas, expressed in years.

Source: Author's calculations based on *Oil & Gas Journal* (2020a, 2020b).

by partner IOCs, which, given their more sophisticated technology, are still more efficient and effective.

The processes and systems involved in extracting, producing, refining, and distributing oil and gas are highly complex and capital-intensive, requiring state-of-the-art technology. Indeed, it could be said that the oil and gas industry performs a modern miracle: in a very short span of time—typically from two to four weeks—it undoes what nature took up to two hundred million years to perform (i.e., it returns the carbon atoms of the hydrocarbon molecules trapped deep underground in sand and rock to the surface, ultimately ending up as carbon dioxide emissions and other harmful pollutants in the atmosphere). Moreover, given the discrepancy between the localization of oil reserves and demand and the fact that carbon atoms are free to cross any borders, the same miracle also performs a spatial redistribution of released carbon atoms (through emissions), meaning they eventually accumulate in the atmospheric global commons. In other words, the oil industry operates as a gigantic multibranched era-hopping conveyor belt, transporting carbon stocks embedded in the earth from the distant past to the current day. But as is widely acknowledged, carbon emissions are not the sole consequence of oil production: the fossil fuel industry—both literally and figuratively—greases the cogs of the global economy.

The oil industry's activities are mainly twofold, divided into so-called upstream operations of exploration and production and downstream operations of refining and distribution. Given the high entry costs, the world's major oil and gas companies are typically integrated (i.e., they carry out both upstream and downstream activities). In brief, exploration includes prospecting as well as seismic and drilling activities that take place before the development of a proper oil field, production involves the extraction of oil from below the ground via onshore and offshore drilling, refining is the process of eliminating unwanted components to obtain clean hydrocarbons that are then used to produce distinctive end products, and finally, in the distribution phase, these products are transported to wherever demand requires through a well-organized system of pipeline networks, seafaring tankers, and global railway and road networks.

Oil industry activities are complex, multifaceted, and painstaking, but with a jaw-dropping $3 trillion of annual revenues (IBISWorld 2020), oil is a highly profitable industry that sees basically a handful of IOCs and NOCs jockeying for position. IOCs are private entities whose business operations traditionally cover the full cycle from exploration through production and refinement to distribution of petroleum products. NOCs are by and large similarly structured, but they are fully or largely owned by a state. Traditionally, IOCs developed as *resource seeking* to supply their downstream activities of refinement and distribution; NOCs were instead considered *market seeking* since they were supposed to look for new markets to distribute their products.

This distinction is no longer tenable, however, for a number of concurrent reasons. First, NOCs no longer operate on the basis of a national political logic and are now equally driven by commercial goals; IOCs and NOCs increasingly cooperate globally in developing more challenging oil fields. Some NOCs, especially from Asia, are active resource seekers in upstream competition with IOCs because their countries of origin do not have oil reserves, while the shrinking European and North American oil markets push IOCs to seek new terrain in the marketplace.

The largest IOCs—such as BP, Chevron, ExxonMobil, Shell, and Total SA (renamed TotalEnergies in June 2021)—are huge multinational, vertically integrated firms based in the United States and Europe with extracting and distribution operations worldwide. IOCs reigned supreme in the oil world until the 1970s, thanks to the long-term concession agreements dating back to the colonial era and maintained in the immediate years after decolonization, in part due to host countries' lack of the technical know-how in exploration and production. IOCs saw this supremacy gradually eroded by the growing role of state-owned NOCs, initially established by and headquartered in the major exporting countries. In the post–World War II period, many oil states began a major campaign to take back control of their own underground reserves, and by the 1970s this process had all but concluded.

The basic principle that led to the creation of NOCs in oil-rich postcolonial states was that of achieving permanent national sovereignty over natural resources, sanctioned in many United Nations declarations, resolutions, and treaties. NOCs, thanks to their ownership of reserves, have developed extensive vertically integrated global networks for the distribution of their oil-based products. Some of the biggest oil companies in the world are NOCs (as tables 1.2. to 1.5 show): CNPC/PetroChina, Russia's Gazprom, National Iranian Oil, and Saudi Arabian Aramco—the world's biggest, with a 10 percent share of crude oil—have a production capacity that can extract up to twelve million barrels of crude out of the ground per day. NOCs, however, are not necessarily structurally identical; a useful distinction is between those belonging to countries that hold large amounts of reserves and are oil exporters and those based in oil-importing countries, typically in Asia. In oil-exporting countries, NOCs were founded as a political response to the perceived traditional exploitation of their oil reserves by IOCs. This process, which started in 1938 with Mexican Pemex, culminated in the 1970s when most Middle Eastern countries as well as some Western ones—Canada, Norway, and the United Kingdom—spurred by strong increases in oil prices, established their NOCs. The oil-poor Asian countries—China, India, and South Korea—established their NOCs in the 1980s and 1990s with the objective of targeting international resources, purchasing new properties, and participating in other oil companies. Given their size, increasing dominance over global reserves and share of global oil production, NOCs' importance has risen significantly in comparison to IOCs, as evinced by tables 1.2, 1.3, 1.4, and 1.5.

THE OIL AND GAS INDUSTRY AND CLIMATE CHANGE: ACKNOWLEDGMENT AND COEXISTENCE

It would be an understatement to say that the relation between the oil industry and climate change remains awkward and controversial. For instance, Mulvey and colleagues investigated the position on climate change of eight major fossil fuel companies, including five IOCs (BP, Chevron, ConocoPhillips, ExxonMobil, and Shell), based on their January 2015 to May 2016 communications, documents, and actions. The study showed that while all the oil companies analyzed openly acknowledge climate science and plan for a less carbon-intensive business model, at the same time they "maintain membership—and in many cases have leadership positions—in trade associations and other industry-affiliated groups that spread disinformation about climate science and/or seek to block climate action" (Mulvey, Allen, and Frumhoff 2016, 2).

Or, to put a specific IOC under the microscope, it took Shell more than sixteen years to caution its shareholders that climate change represented a financial calamity for the company, despite having privately known for decades about the causality relation between its products and climate change, as chapter 2 will show. At the same time, in 2018 Shell began lobbying the US Congress to introduce a carbon tax. Meanwhile, the Texas oil industry—ironically—is expecting the government to cough up taxpayer money to pay for a sixty-mile-long seawall to protect its refineries in the Gulf of Mexico from the more powerful storms and higher tides that climate change is causing.

The coexistence between the oil industry and climate change has been challenging since the latter became an issue, with the industry often partaking in duplicitous behavior with regard to environmental concerns. Some serious science took place in the research facilities of the oil companies themselves, leading them to conclude incontrovertibly that climate change was happening and that it was real and dangerous. Nonetheless, they continued to pour cash into bogus scientific research and think tanks, with the precise aim of sowing the seeds of doubt over the very same evidence that their internal scientists had produced.

It is usually assumed that anthropogenic climate change became part of the wider public discourse after the 1990 first assessment report of the International Panel on Climate Change (IPCC), *Climate Change: The IPCC Scientific Assessment* (IPCC 1990). A more prudent benchmark of awareness can be set at 1992: in that year during the Rio Conference, heads of state and delegates were officially informed about the global scientific consensus on the harmful effects of greenhouse gas emissions (presented in a supplementary assessment report of the IPCC, *Climate Change: The IPCC 1990 and 1992 Assessments* [IPCC 1992]). Since that point in time, any claims of obliviousness about the consequences of emissions and the alleged impotence of oil and gas companies to reduce their contribution have become inexcusable.

In truth, the oil industry had already discovered the nexus between their produce and climate change decades earlier (see chapter 2). Scientists of Humble Oil in the United States (which was later absorbed into Standard Oil, eventually evolving into ExxonMobil) published research acknowledging the science of climate change in peer-reviewed journals from as early as 1957. From 1968 onward, these warnings were reiterated to the oil industry even in the dire terms that have today become overly familiar: melting ice caps, rising sea levels, more intense and frequent extreme events, and serious environmental damage on a global scale. In Europe, Shell knew too: internal documents circulated in the 1980s attest that the company acknowledged the seriousness of climate change and that Shell's products were responsible for it.

At any rate, at the dawn of climate policy in the early 1990s, IOCs point-blank refused to modify their business model in order to mitigate global warming, in some cases viewing the mounting pressure to curb green-house gas emissions as a conspiracy to disrupt the industry and the status quo of their business models. The majority of NOCs, on the other hand, shielded by more protective governments (by and large in less democratic societies) and not subjected to public criticism or opinions that tend to be either stifled or provoke less public indignation, seemingly ignored climate change until a few years ago, when some of them—CNPC/Petro-China, Pemex, Petrobras, Statoil (now Equinor), and Saudi Aramco—eventually joined the *Oil and Gas Climate Initiative*, a voluntary alliance focused on leading the oil industry response to climate change.

The basic reasoning behind the IOCs' position was that any kind of limit on emissions would directly threaten their revenues and profits; they were also quick to point out how the global industrial stage depends on their products and how any curbs on them would have adverse knock-on effects on the world economy. To defend their business model, the major IOCs—Amoco, BP, Chevron, Exxon, Mobil (since 1999 the latter two have merged into ExxonMobil), Shell, and Texaco—used the *Global Climate Coalition*, an advocacy group of businesses put together with the help of public relations giant Burson-Marsteller in 1989, just one year after the first IPCC report, to promote climate denial. Through the Global Climate Coalition, the biggest IOCs cast doubt on the science of climate change and opposed policies against emission cuts.

In 1996, BP left the Global Climate Coalition. The following year BP publicly broke ranks with its still obstinately skeptical American peers and called for a precautionary approach to climate change: in 1997 BP's

CEO, John Browne, acknowledged the connection between greenhouse gas emissions and climate change, pledging that BP would help tackle the problem by shifting to a less carbon-intensive business model. Shortly afterward Shell followed suit. This was the dawn of the *Atlantic divide* between major IOCs, in particular between ExxonMobil and BP, as vividly illustrated by Lovell's (2010, 42–66) report of a 2003 debate between Frank Sprow and Greg Coleman, senior representatives from ExxonMobil and BP, respectively, on the responsibility of oil companies in the face of climate change.

The responses of the major IOCs in the United States were surprisingly different until approximately the first few years of the third millennium. In brief, US companies Chevron and, in particular, ExxonMobil stood firm in their denial of anthropogenic climate change, proclaiming the ruinous cost of greenhouse gas control as they simultaneously lobbied against climate policy and invested very little in alternative sources.

The European majors BP and Shell, on the contrary, accepted the scientific basis of anthropogenic climate change and espoused the principle of precautionary action, making declarations in support of the Kyoto Protocol and pledging substantial investments in renewables. Powerful corporations have the capacity and lobbying influence to shape environmental policy, as opposed to merely voicing support or opposition, but in this instance US IOCs adopted a markedly reactive stance based on the rebuttal of responsibility for climate change, while their European counterparts embraced a proactive strategy acknowledging a degree of accountability.

ExxonMobil's response to climate change was a particularly long and inconsistent process. It was prompted, paradoxically, by the Rockefeller Foundation—costarted by John D. Rockefeller Sr., the founder of Standard Oil—that from 2004 pressed the company through letters, meetings, and shareholder resolutions to acknowledge climate change, abandon climate denial, and direct its business model toward clean energy. In 2007, ExxonMobil disclosed to shareholders, albeit in somewhat ambiguous language, the financial risks to profitability of climate change, all the while continuing to fund climate denial; the 2008 report presented at the annual general meeting pledged to stop pouring company resources into campaigns to deny climate change. Only in April 2014 did ExxonMobil publish a report publicly acknowledging climate change for the first time.

Unsurprisingly, however, a 2017 study by Supran and Oreskes found that until 2014 the oil giant had systematically misled the public about climate change; the study basically argued that while ExxonMobil's peer-reviewed scientific publications acknowledged the scientific consensus on climate change, internal documents and paid editorial-style advertisements (*advertorials*) in major newspapers denied it. The more the latter group of documents were aimed at the public, the more they were steeped in skepticism. Private correspondence acknowledged the scientific consensus, whereas openly available statements espoused climate denial: "We find that as documents become more publicly accessible, they increasingly communicate doubt" (Supran and Oreskes 2017, 1), highlighting those advertisements as being particularly effective in that regard. Ironically, ExxonMobil produced valuable climate science: "83% of peer-reviewed papers and 80% of internal documents acknowledge that climate change is real and human-caused" but on the other hand stated in its public pronouncements that "only 12% of advertorials do so, with 81% instead expressing doubt" (Supran and Oreskes 2017, 1).

Nowadays, attitudes and intentions seem—at least to the casual observer—to be inching their way toward change. All the largest IOCs have acknowledged that anthropogenic climate change is real and claim that a low-carbon future is somewhere on their horizons. Even ExxonMobil, possibly the most obstinate climate opponent, states that the Paris Agreement is "an important step forward by world governments in addressing the serious risks of climate change" and concedes that "the company has a constructive role to play in developing solutions" (ExxonMobil 2016) Actions, of course, speak louder than words, and as carefully testified by Supran and Oreskes (2017) and underlined in chapter 2, ExxonMobil's attitude toward climate change remains highly ambiguous. Similarly, other IOCs still need to account for some lack of clarity in their conduct.

Oil and gas companies envisage different courses of action for a low-carbon future, from investing in renewables to modifying their business models in such a way that would see them limiting their exploitation of the oil and gas reserves they hold, all the while promoting and employing carbon removal technologies. The rationale behind Big Oil's willingness to establish a new code of conduct might be explained away by an obstinate instinct for survival: if oil and gas companies want to maintain their social license to operate in a climate-endangered world, they must modify their outlook and, as a consequence, their operations in accordance with the mounting pressure and the nascent social/moral norms that aim at delegitimizing the wealthy Westernized carbon-intensive lifestyle model; the same process occurred with other socioeconomic practices that were once deeply entrenched and influential, such as slavery and tobacco. As the title of this book suggests, could the near future bring the spectacle of seeing Big Oil shed its villainous guise to become a global paradigm for green energy? Most oil majors openly declare their willingness to aim at this objective; their meaningfulness has yet to be proved.

Other circumstances exist that could advance the cause of decarbonizing the oil industry, the first one being the boom in production of renewable energy in recent years with its resulting drop in prices (IEA 2020b, 2020c). Second, in the last seven years oil prices fluctuated from their 2014 high of $100 per barrel to lows of $27, even floundering below zero in April 2020, a crippling price caused by the effects of the COVID-19 pandemic. This was a serious blow to the economic certainties of producers, caused not only by the pandemic but also by the limited capacity to store the oversupply of oil in addition to trade tensions, political ambiguities, and the short-term inelasticity of supply and demand.

At the same time, the debate on climate change and civil society has thus far paid little attention to NOCs despite their importance in the present-day oil world, as emphasized above. By the same token, they themselves have not been particularly reactive to the challenges posed by climate change, since NOCs are subject to far fewer pressures than IOCs, often typecast as the *"pantomime villains"* in the global climate discourse.

However, as a result of the Paris Agreement, the involvement of NOCs in climate change action is set to substantially increase. Under this agreement, countries voluntarily make emission reduction commitments (known as *nationally determined contributions*) that generally entail ambitious regulatory and policy changes. It is likely that countries with *national champions* in the oil business delegate to them the bulk of the effort to reduce emissions since, those companies being state-owned, this choice better testifies to the host countries' genuine involvement in the endeavor. However, NOCs seem less prepared than IOCs to face the challenges posed by climate regulations and policy, since they have traditionally been given very ample room for maneuver by their respective governments. Whereas IOCs—long used to competing in difficult markets—have a wide range of options for decarbonizing their business, from renewables to carbon capture and storage technologies, NOCs seem to be facing a more limited menu, a more demanding one, because, as made clear by their ownership (table 1.1), these options would largely depend on the governments running the countries; their very nature means they will have a much broader focus than any individual company.

Action against climate change by NOCs could be favored by the Organization of the Petroleum Exporting Countries (OPEC), a permanent intergovernmental organization of fourteen oil-exporting countries founded in 1960 that coordinates and unifies the petroleum policies of its member countries (all of them have NOCs, some of which are among the largest in the oil and gas industry). The OPEC secretary general, in a speech at the 2017 International Petroleum Week in London, reiterated the fact that the organization is committed to tackling climate change, as demanded by the Paris Agreement, through support for a shift to renewables by its member countries and therefore NOCs. A further glimmer of hope in regard to the capacity of NOCs to adapt their business to less carbon-intensive models is provided by membership of some of the largest NOCs (Equinor, Pemex, Petrobras, Saudi Aramco, and CNPC/PetroChina) in the Oil and Gas Climate Initiative, made up of thirteen member companies from the industry, including giants such as Shell and Exxon, and established in 2014 with the precise intent of reducing dangerous greenhouse gas emissions.

THE OIL AND GAS INDUSTRY'S DIRECT CONTRIBUTION
TO CLIMATE CHANGE: GREENHOUSE GAS EMISSIONS

The most straightforward testament to the role the oil and gas industry has played—and still plays—in the climate crisis is the direct contribution in terms of greenhouse gas emissions generated by its oil-related activities.

Research by Richard Heede's Climate Accountability Institute—as well as a number of other studies—has focused on the contribution of the large carbon producers to global cumulative emissions of major greenhouse gases, such as carbon dioxide and methane. Perhaps their most remarkable finding is that 62 percent of the global industrial emissions of carbon dioxide and methane from 1751 to 2015 can be traced to the activities of one hundred currently operating carbon majors (forty-one public investor–owned companies, sixteen private investor–owned, thirty-six state-owned, and seven government-run) and eight nonextant ones. The emissions traced to carbon majors are calculated based on the carbon content of fuels marketed (subtracting nonenergy uses); carbon dioxide from cement production as well as from flaring, venting, and own fuel use; and fugitive or vented methane (Heede 2014). Heede's data also demonstrates that given the rapid global industrialization of the last few decades, the one hundred currently operating carbon majors have produced 71 percent of global industrial emissions since 1988; the top emitters are fossil fuel corporations (oil and gas as well as coal companies), with cement producers making up a small minority. The original 2014 database, for instance, included only seven cement producers whose emissions amounted to 1.45 percent of carbon majors cumulative total (Heede 2013, table 4).

Moreover, a study published in 2017 by Ekwurzel and colleagues extends Heede's 2014 conclusion by linking carbon majors' fossil fuel–related activities to atmospheric carbon dioxide and methane concentrations as well as to relevant climate impacts, namely the global mean surface temperature (GMST) and the global sea level (GSL), with the latter widely recognized as being one of the major consequences of climate change. Strikingly, Ekwurzel and colleagues' study (2017, 579) found that the historical (1880–2010) and recent (1980–2010) emissions of ninety major carbon producers resulted in "~57% of the observed rise in atmospheric CO_2, ~42–50% of the rise in GMST and ~26–32% of GSL rise over the historical period of 1880–2010 and ~43% (atmospheric CO_2), ~29–35% (GMST), and ~11–14% (GSL) since 1980." In the same vein, a 2019 study by Licker and colleagues showed the nexus between the drop in surface ocean pH levels and carbon production, demonstrating that eighty-eight of the carbon majors were responsible for 55 percent of the acidification of the oceans between 1880 and 2015, with as yet inestimable damage to ecosystems and marine life not to mention the fishing industry, so vital to myriad coastal communities (Licker et al. 2019).

Importantly, carbon majors have produced more than half of their emissions roughly in the past thirty years, when the global community was already well aware of the potential dangers of climate change; 833 gigatonnes (Gt) of carbon dioxide (50.4 percent) of the emissions associated with carbon majors' activities have been produced since 1988, whereas 820 (49.6 percent) were produced in the period between 1750 and 1987 (CDP 2017). More generally, as Heede (2014, 234) claims, "Of the emissions traced to carbon major fossil fuel and cement production, half has been emitted since 1986."

Also of note is that the industry seems to have had a far bigger direct contribution to climate change through methane emissions associated with the extraction of fossil fuels: since the industrial revolution, extraction processes have released 25–40 percent more climate-changing methane—an increase determined mostly by unaccounted flaring and venting, underreported accidents and leaks, and the expansion of fracking activity—than previously thought (Hmiel et al. 2020).

With specific regard to Big Oil, its contribution to global greenhouse gas emissions is in many respects striking. The top ten companies in terms of cumulative emissions of Heede's 2014 study all belong to the oil and gas industry. The biggest sixty oil and gas companies contributed to more than 40 percent of cumulative global industrial emissions in the period 1988–2015, the top ten accounted for almost 22 percent, and the top twenty were responsible for more than 30 percent, as shown by table 1.6. The oil and gas industry holds fossil fuel reserves that if burned would bring the planet well above the 1.5°C warming target, the temperature beyond which the most severe climate impacts hit. A study by Welsby et al. (2021) stated that, in order to have a 50 percent chance of limiting global heating to that threshold, no more than 40 percent of current oil and gas reserves must be extracted between now and 2050.

These figures and considerations give an idea of the salience of greenhouse gas emissions by oil and gas companies in climate change. A fundamental clarification is in order here: by indiscriminately providing their products to the global economy, these companies are the heartbeat of the current carbon-intensive socioeconomic system. Their prominent role in the climate crisis and the important implications for climate change as well as the sustainability discourse should place these companies at the center of the climate debate. By and large, states are the principal players involved in addressing climate change. Other stakeholders, such as civil society, individuals, local authorities and communities, private-sector actors, and international institutions are considered subordinate players. While all stakeholders are to different extents involved in global efforts to combat climate change, oil and gas companies are, in relation to their actual prominence, the truly overlooked player in the current climate policies and initiatives.

Considering how strongly these particular corporate entities are implicated in contributing to and perpetuating the climate crisis, it is unacceptable to equate their position with that of the business world in general or indeed of other stakeholders. Oil and gas companies have a very distinct, specific, and crucial role in the climate issue, Considering the extent to which they contributed to the problem, their power and wealth, the benefits they derive from their fossil fuel–related activities, and their technical expertise, at best it is irrational to view them merely as subordinate players in global climate governance, and at worst it is preposterous. Of course, they are subject to the binding emission limits imposed by the national and subnational political authorities, and similar to other corporations outside the carbon business, oil and gas

Table 1.6

Oil and gas companies' scope 1+3 greenhouse gas emissions 1988–2015, GtCO$_2$e and percent of global industrial emissions 1988–2015

Oil and Gas Company	Emissions	Percentage	Typology
Saudi Aramco	40.6	4.5%	NOC
Gazprom (Russia)	35.2	3.9%	NOC
National Iranian Oil	20.5	2.3%	NOC
ExxonMobil (USA)	17.8	2.0%	IOC
Pemex (Mexico)	16.8	1.9%	NOC
Shell (UK/Netherlands)	15.0	1.7%	IOC
CNPC/PetroChina	14.0	1.6%	NOC
BP (UK)	13.8	1.5%	IOC
Chevron (USA)	11.8	1.3%	IOC
PDVSA (Venezuela)	11.0	1.2%	NOC
Abu Dhabi National Oil	10.8	1.2%	NOC
Sonatrach (Algeria)	9.0	1.0%	NOC
Kuwait Petroleum	9.0	1.0%	NOC
TotalEnergies (France)	8.5	0.9%	IOC
ConocoPhillips (USA)	7.5	0.8%	IOC
Petrobras (Brazil)	6.9	0.8%	NOC
Lukoil (Russia)	6.7	0.8%	IOC
Nigerian National Petroleum Corp	6.5	0.7%	NOC
Petronas (Malaysia)	6.2	0.7%	NOC
Rosneft (Russia)	5.9	0.7%	NOC
Total 20 (Top 10)	**273.6 (196.6)**	**30.4% (21.9%)**	

Source: Elaboration from the Carbon Majors Database—2017 Dataset Release (CDP 2017). According to the Greenhouse Gas Protocol of the World Resources Institute (WRI n.d.), *scope 1* emissions refer to direct oil and gas combustions, and *scope 3* emissions originate from the downstream combustion (for energy and nonenergy purposes) of oil and gas that they have distributed within the global economic system. Indeed, the largest share (roughly 90%) of oil companies' emissions consists of scope 3 emissions.

companies voluntarily disclose their greenhouse gas emissions and integrate effective abatement strategies into their business models. This is the case, for instance, of the *Carbon Disclosure Project* and, in relation to methane emissions, of the Climate and Clean Air Coalition's *Oil & Gas Methane Partnership*. However, given the nature of their core business, this is far from adequate.

The oil and gas industry has had a unique role in causing, shaping, advancing, and defending the current unsustainable fossil fuel–dependent global economy and for decades has been dictating the rules of the game in terms of the world's reliance on oil. Through its informed and self-advantageous choice to continue the exploration, production, refinement, and distribution of oil and gas after the 1990s—all the time denying the harmfulness of such products and using its lobbying clout on political decision makers—Big Oil has imposed this reliance on fossil fuels on other industries, which have had to shape their business models accordingly with a limited number of costly alternative options; the same is true for individuals, whose lifestyles have evolved in parallel to the business choices made by this influential and formidable industry.

Recognition of the prominent role of oil and gas companies in causing and perpetuating climate change does not imply that they should be center stage in addressing the issue. Different agents have different roles and responsibilities in tackling climate change, first and foremost states, which should provide the appropriate legislative and political frameworks for ensuring that in accordance with their responsibility, oil and gas companies comply with their duties. A hybrid multilateralism should emerge, with Big Oil's role in it fitting to the role that the industry played in climate change, with states, individuals, and other agents sharing the stage.

In sum, fossil fuels should now be looked upon as a harmful product, the use of which is affecting the health, lives, and well-being of present generations of all Earth's inhabitants and will continue to do so in the future. As was the case with industries dealing with products such as tobacco, asbestos, and lead—once admissible but later banned or reviled on the basis of sound scientific evidence of their harmfulness—it is time to acknowledge not only the role of the oil and gas industry but also the moral and political implications deriving from its involvement in such harmful products.

The oil industry must develop a viable vision for a transition to a low-carbon future if it still wants to be an active part of it. Step one is for Big Oil to admit, perhaps in the first instance to itself, that the old world is unquestionably mutating and that the new low-carbon world will not—and should not—forget its role in endangering the old one. This implies that the contributions that oil and gas companies have made to climate change give rise to their responsibility; this responsibility, in turn, means that these companies have duties (i.e., standards of moral behavior inspired by principles of justice that involve a practical commitment) for, quite simply, doing or refraining from doing something.

At any rate, a paradigm shift is under way, and to be a part of it, Big Oil must ensure that it no longer abrogates its responsibility and duties by contributing—under the essential stewardship of other stakeholders—to drawing up a concrete road map to illustrate change.

By indiscriminately flooding the global economy with fossil fuels, oil and gas companies are the driving force behind the current carbon-intensive socioeconomic system. Yet they have always somehow managed to narrowly escape condemnation, indeed in many cases even recognition, for their role in that system. Moreover, besides their direct contribution of greenhouse gases in the atmosphere, some fossil fuel majors over the years have spared no expense in oiling the climate change denial machine through funding and lobbying, despite being fully aware of the perils associated with climate change, and have taken no steps to modify their extremely profitable business to meet the challenges that Earth is facing.

Big Oil's direct contribution to climate change through carbon emissions establishes its causal responsibility, a necessary condition for more stringent notions of responsibility but one that alone is not sufficient to justify consequent compelling duties.

To establish and justify the more stringent moral responsibility and consequent duties that Big Oil must shoulder in a pluralistic and nonarbitrary way, a solid morally relevant factual basis must form the framework. As historian of science Naomi Oreskes (2019) argues, in order to command the authority required to engage agents who should accomplish the necessary transformations, climate change discussions need to be based on facts.

The morally relevant facts for the oil industry generally relate to harm; in this regard, agents may be considered responsible if they are aware and/or are able to foresee that their action(s) bring about harm and if they have the capacity and possibility, but lack the willingness, to avoid or minimize harm (Hart 1963). These specifications underpin a classification system to establish the morally relevant facts.

For ease of reference, these facts can be subdivided into five groups, as table 2.1 shows: *fact A, awareness*: long before the 1990s heralded a more widespread understanding of the issue, oil companies knew that their fossil fuel–related activities provoked dangerous climate change; *fact B, behavior*: oil companies have not changed their fossil fuel–centered behavior; *fact C, capacity*: less carbon-intensive alternatives were possible; and *fact D, denial*: through denial campaigns, major international oil companies (IOCs) successfully opposed political efforts to decarbonize economic systems and to act on the climate change already under way. The fifth, *fact E, enrichment*, will be examined shortly.

These morally relevant facts suffice to form the consensus that oil companies *enabled* harm (Foot 1967) to humanity and the planet, as opposed to their direct contributions in terms of emissions that *did* harm (and on this basis generated their causal responsibility). In brief, harm-enabling morally relevant facts involve the removal of obstacles that prevent harm

Table 2.1

The morally relevant facts

Fact A	Awareness	The industry was aware of the damage its products provoked to the climate.
Fact B	Behavior	The industry did not modify its behavior.
Fact C	Capacity	Less carbon-intensive products were possible.
Fact D	Denial	Oil majors mounted huge denial campaigns.
Fact E	Enrichment	Fossil fuel companies derived staggering profits from their harmful activities.

Source: Author.

or hinder actions that prevent harm (Barry and Øverland 2016), as examined in more detail in chapter 4. Oil companies actively obstructed the recognition of their activities is an aggravating factor in climate change; the latter is an essential prerequisite before steps to reduce harmful activities can be taken. The consequent assumption is that the moral cogency of doing and enabling harm is the same; that is, *doing* harm is morally equivalent to *enabling* harm in terms of responsibility, since both are contributing factors to climate change.

Furthermore, there is a stand-alone fact (the fifth, *fact E, enrichment*) that represents and embodies the *raison d'être* of oil and gas companies' business mission: the staggering profits deriving from their fossil fuel–related activities make this fact significant in moral terms. Despite not being intrinsically wrong from a moral standpoint in that it is unrelated to harm and therefore does not concur to climate change, fact E represents a distinctive and complementary moral basis—justified through the *beneficiary pays* and *ability to pay* moral principles of climate ethics, as specified in what follows—for determining and more effectively defining oil companies' moral responsibility for climate change. So, despite its irrelevance in terms of harm, fact E is incorporated into this chapter.

FACT A: AWARENESS

Oil and gas companies were aware of the threats of climate change, but they sequestered this knowledge away from shareholders, stakeholders, and the general public. Some of the major IOCs, such as ExxonMobil and Shell, had high-level internal scientific and technical expertise and were aware of the available scientific knowledge about potential harmful effects for the global climate system—especially in terms of atmospheric temperature increase—of burning fossil fuels (CIEL 2017; Franta 2018). Since the 1970s at least, the oil industry even knew that air pollution from fossil fuel combustion posed serious risks to human health, albeit deliberately casting doubt on the issue (Milman 2021); in 2018 its products caused 8.7 million deaths, that is, 1 in 5 of all deaths worldwide (Vohra et al. 2021).

By enacting this knowledge concealment, the oil industry prevented other subjects from better grasping the nature of climate change and thereby taking action against it based on this extensive and in-depth knowledge of its causes and dynamics.

The year 1990 marked a turning point in terms of the general awareness of the perils of anthropogenic climate change when the International Panel on Climate Change's (IPCC) first assessment report was published (IPCC 1990), revealing the global scientific consensus on the issue. But knowledge of the potentially negative consequences of carbon emissions on the planet dates even further back to the nineteenth century and was widespread among different scientific communities. Oil companies too had already known about climate change for decades, possibly even since the inception of the industry. The threat posed by the ever-increasing carbon emissions was initially underestimated given the belief that the oceans would have safely absorbed them, thus eliminating their danger to the climate system. As far back as 1938, however, at least one scientist (Callendar 1938) measured a noticeable impact of CO_2 emissions on global temperatures (0.005°C per year for the previous fifty years), evidence eventually confirmed—and thereafter referred to as the *Suess effect*—by the Scripps Institute of Oceanography's chemist Hans Suess (1955). In 1957 a landmark work by Revelle and Suess demonstrated unequivocally that not only would the world's oceans *not* absorb CO_2 as rapidly as previously imagined but also that its level in the atmosphere was likely to increase significantly (Revelle and Suess 1957). Two months later scientists at Humble Oil (a subsidiary of Standard Oil New Jersey, now ExxonMobil) submitted their findings on the same topic, which similarly recognized the increase in atmospheric CO_2 and acknowledged the connection between fossil fuel combustion and said increase as well as the link between atmospheric CO_2 and potential temperature increases (Brannon et al. 1957).

From the 1940s, the Western oil industry began carrying out groundbreaking research into climate change and its impacts. The research focused on long-term changes in Earth's temperature, the relationship between global temperatures and sea level rise, changes in the concentration of CO_2 in the atmosphere, and the nature, causes, and history of hurricanes and even explored the techniques, technologies, and consequences of intentional weather modification (CIEL 2017).

By the late 1950s, the North American oil industry (and very likely European IOCs too, given the highly oligopolistic structure of the industry at the time) was involved in research on the accumulation of CO_2 in the atmosphere and on the contribution of the combustion of fossil fuels to such phenomenon through the American Petroleum Institute (API), the US oil and gas industry's trade association. The API's Smoke and Fumes Committee's main objective was to combine industry-funded research—usually undertaken to prove a predetermined result, according to "credible firsthand accounts" (CIEL 2017, 21)—and public relations advocacy in order to increase skepticism about air pollution science with the ultimate goal of swaying legislation on critical issues related to CO_2, among others. One such example is a 1958 project aimed at measuring the Suess effect (i.e., the proportion of atmospheric carbon of fossil origin).

At the same time, at the one hundredth anniversary celebration of the oil industry in the United States in 1959, organized by the API in New York, the renowned physicist Edward Teller warned oil executives, government officials, and scientists with startling prescience about the correlation between carbon dioxide and global warming. A pattern was forming that would repeat itself over the following decades.

For example, in 1968 the Stanford Research Institute presented the API with a report titled *Sources, Abundance, and Fate of Gaseous Atmospheric Pollutants* (Robinson and Robbins 1968) that summarized the causes, nature, and consequences of global warming and climate change. The report concluded, at page 109, that fossil fuel combustion was the most likely cause for climate change and that climate change could have major impacts worldwide and advocated that the industry should channel significant resources into funding technologies for reducing emissions. The report did not advance definitive claims on climate change but did state that "significant temperature changes are almost certain to occur by the year 2000, and these could bring about climatic changes." In short, damning evidence exists that by 1968 the API—and therefore the American oil industry—knew about the relation between fossil fuel combustion and rising atmospheric CO_2 concentrations, with the consequential temperature rise, and were aware of the need to research means for addressing and controlling CO_2 emissions from fossil fuel combustion. In 1969, the API asked the Stanford Research Institute to better substantiate its original findings. The submitted supplementary report (Robinson and Robbins 1969) reiterated, in its section on CO_2, the conclusions of the 1968 work and stressed that atmospheric concentrations of CO_2 were increasing and that 90 percent of this rise could be attributed to fossil fuel combustion. The report went on to surmise that continued use of fossil fuels would inevitably result in even greater CO_2 concentrations in the atmosphere.

In 1972, the US National Petroleum Council—an advisory committee under the US Department of Energy that advises the federal government on questions related to the oil industry—submitted a report to the US Department of the Interior that basically acknowledged the findings of the 1968 and 1969 Robinson and Robbins reports, albeit presenting the relationship of fossil fuel combustion and CO_2 concentrations to temperature increase in more ambiguous terms.

In the 1970s, in-house research teams from major oil companies informed executives of the consequences of fossil fuel combustion; on various occasions Exxon internal memorandums detail how the company's own scientists alerted management about the correlation between fossil fuel combustion and climate change as well as the imperative of taking serious action against it. In 1978 in one of these internal memos, Exxon senior scientist James Black was categorical about the urgency of the climate risk generated by burning fossil fuels: "Present thinking holds that man has a time window of five to ten years before the need for hard decisions regarding changes in energy strategies might become critical" (Black-Kalinsky 2016).

By the early 1980s, Exxon internally acknowledged that an increase of CO_2 concentrations in the atmosphere due to fossil fuel combustion was wreaking havoc on the climate—especially in terms of rising temperatures. In 1981 Roger Cohen, director of Exxon's Theoretical and Mathematical Sciences Laboratory, was crystal clear in a communication to Exxon's Office of Science and Technology: "A clear scientific consensus has emerged regarding the expected climatic effects of increased atmospheric CO_2"; the communication concluded by claiming that "the results of our research are in accord with the scientific consensus on the effect of increased atmospheric CO_2 on climate" (Cohen 1982). Exxon's Environmental Affairs Program hastened to inform the company's executives on and familiarize them with the climate change debate in the very same year.

It is difficult to imagine, though, that such knowledge remained within the confines of the United States, even in the preglobalization of information era. Indeed, on the other side of the Atlantic, Shell's grasp of climate change intelligence has been documented, specifically in a number of internal documents drafted from 1981—well before the 1990 global scientific consensus on the negative effects of anthropogenic climate change—in which the Anglo-Dutch oil giant recognized that unabated carbon emissions could lead to a series of effects: an increase of between 1.5° and 3.5°C of atmospheric warming, major social and economic upheavals, and severe environmental damage, including the disappearance of entire ecosystems. Shell acknowledged that carbon emissions largely originated from the combustion of fossil fuels and that all its fossil products significantly contributed to the problem (Small and Farand 2018). By the same token, in a 1988 confidential document titled *The Greenhouse Effect*, Shell admitted that climate change could lead to large-scale forced migration, especially due to crop failure and extreme weather modifications in more sensitive regions. In 1991, Shell even produced a film for public release: *Climate of Concern*. It cut to the chase, openly asserting that climate was changing faster than at any time since the last ice age and that this would have worrying impacts on the planet and its inhabitants. Yet the company continued to develop future scenarios largely reliant on oil, publicly stating that fossil fuels were the only realistic way to achieve sustainable development.

Between 1979 and 1983, the API established a task force to monitor and share research on climate change among its members. Notably, members included representatives from almost every Western IOC: Exxon, Mobil, Amoco, Phillips, Texaco, Shell, Sunoco, Standard Oil of Ohio and of California, and Gulf Oil, the predecessor to Chevron (Banerjee 2015).

The oil world, despite the enormity of its main actors, is a small one where critical information spreads like wildfire. As Exxon and Shell's very own scientists were making clear that they knew about the knock-on effect of fossil fuel emissions, it is hard to conceive that the rest of the industry—including national oil companies (NOCs)—were not in the loop. In short, it seems safe to claim that oil majors have known for several decades that their activities were causing long-term damage to the climate.

FACT B: BEHAVIOR

It is extremely difficult to analyze oil companies' behavior in relation to climate change due to the often duplicitous attitude they have demonstrated toward it. As pointed out in the previous section, they carried out serious scientific research into climate change and concluded that it was real while at the same time refuting their in-house evidence by not taking action against it and actually denying it.

In the early 1990s, the social and political pressure to act against climate change started to gain momentum, but Big Oil, by and large, did not change its carbon-centered business model. In public, IOCs mostly dismissed the scientific evidence on the relations between fossil fuels and climate change as a leftist attack on the oil world. NOCs, the oil champions of some oil-rich and oil-thirsty countries, seemingly ignored the issue and carried on unconcerned, business as usual. Cutting emissions was seen by the industry as a threat to its very survival, not to mention the domino effect it would have on the many industries contingent on fossil fuels and therefore the general global economy and humankind.

However, the dawn of the new millennium appeared to mark a gradual divergence in attitudes toward the climate crisis, despite the oil majors remaining hesitant about fully engaging in it; in a somewhat rough schematization, at the beginning of the 2000s US IOCs adopted a reactive strategy based on the rebuttal of responsibility for climate change, whereas European IOCs embraced a more proactive approach that accepted some forms of responsibility (Sæverud and Skjærseth 2007), by then conceding that climate change exists and stating that a low-carbon future was one of their goals.

Laudable though this may be, the issue remains that for decades after their internal knowledge—and at least for one decade after public scientific consensus on climate change—oil majors did not switch to less carbon-intensive business models. On the contrary, they continued to explore, produce, refine, and distribute fossil fuels with the same cavalier attitude they had when climate change was just a niche topic on the lips of the few. Exxon, for instance, in the decade immediately following the knowledge accrued and the agreed scientific consensus on climate change in the 1990s, increased its investment in fossil fuels, as evinced by figure 2.1.

Taking into consideration more recent years, after the oil industry had publicly acknowledged climate change and announced its intention to

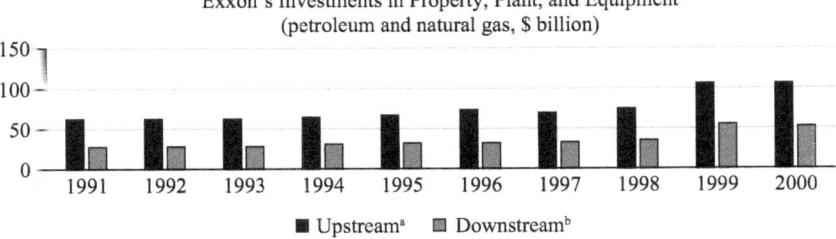

Exxon's Investments in Property, Plant, and Equipment
(petroleum and natural gas, $ billion)

Figure 2.1

Exxon's investments in property, plant, and equipment (petroleum and natural gas, $ billion). [a]Upstream investments include oil and gas exploration and production; [b]downstream investments include oil and gas refining and marketing. *Source*: Exxon 10K *Annual Reports* to the US Securities and Exchange Commission (various years). http://ir.exxonmobil.com/phoenix.zhtml?c=115024&p=irol-sec.

transition to more sustainable business models, it is revealing to scruti-
nize Anglo-Dutch Shell and British BP's budgets. Shell's capital expenditures, or *capex* (i.e., the money it spends to buy, maintain, or improve
its fixed assets for exploring producing, refining, and distributing oil and
gas), remained almost constant in the five (prepandemic) years between
2015 and 2019, with a maximum in 2015 of $26.1 billion and a minimum
in 2017 of $20.9 billion. In the same period BP posted a capex that varied
from a minimum of $17.4 billion in 2016 to a maximum of $25.1 billion
in 2018.[1]

The tableau of the immediate future—in a world where the current
lion's share of energy investments is still in fossil fuels (IEA 2019a), just
as it was thirty years ago—is a *spot-the-difference* challenge with the one
of the past. ExxonMobil stated that the world needs more energy and in
2018, accordingly, announced a drive worth $200 billion in major oil and
gas projects around the world over seven years. A case in point, in March
2019 ExxonMobil announced capital outlays of $32 billion through the
end of 2020, a 24 percent increase from 2018, raising its 2025 profit-
growth target by five percentage points to 140 percent compared with
2017 levels (Crowley 2019). Objectives were to be achieved through an
almost exclusively fossil fuel–centered strategy that greatly speculates on
shale: the company itself announced, for example, that the output of the
cornucopian Permian basin should skyrocket to 1 million barrels per day
by 2024 (ExxonMobil 2019a). In March 2021, ExxonMobil unveiled its
intention to produce 3.7 million barrels a day by 2025 (Crowley 2021),
despite the pledges of its December 2020 emissions reduction plan—
heavily criticized for being a *nonreduction* plan—that had forecast the
production of 1 million barrels per day between 2021 and 2025 to which
eventually 800,000 barrels per day in 2025, produced in a sensitive
marine ecosystem in Guyana that will become the company's largest
single source of fossil fuel in the world (Juhasz 2021), must be added.

A report published by the Global Gas and Oil Network (GGON 2019)
showed that in the four-year period between 2020 and 2024, the industry as a whole plans to invest $1.4 trillion in extraction projects alone.

In Europe, the oil and gas industry maintains that it will continue the production of fossil fuels, with the Shell CEO claiming in an interview with Reuters news agency that "despite what a lot of activists say, it is entirely legitimate to invest in oil and gas because the world demands it. . . . We have no choice but to invest in long-life [fossil-based] projects" (Bousso and Zhdannikov 2019). In 2017, French TotalEnergies signed a multibillion-dollar agreement to develop part of the Persian Gulf South Pars, the world's largest gas field, jointly owned by Iran and Qatar.

In sum, the five largest IOCs—BP, Chevron, ExxonMobil, Shell, and TotalEnergies—will invest around $3.5 billion (only 3 percent of their 2019 capex) in low-carbon technologies, while roughly $110.5 billion will be put into oil and gas exploration and production (InfluenceMap 2019). Paradoxically, at the same time major IOCs—for instance, BP, Shell, Italian ENI, and Spanish Repsol—made a series of pledges, plans, and press releases aimed at clarifying their commitments to achieve net-zero emissions by 2050 (CTI 2020a).

Similarly, NOCs are far from shifting into reverse gear as far as fossil fuel investments are concerned. Saudi Aramco plans to invest $300 billion over ten years in upstream oil and gas, while Russian Gazprom's investment program for 2018 amounted to over $20 billion, largely centered on the development of natural gas projects as well as on the realization of gas facilities and infrastructure projects. Gazprom's oil arm Gazprom Neft will spend roughly $7 billion on developing new oil fields and on the modernization of refineries. Resource-seeking CNPC/PetroChina invested $1.2 billion to buy 10 percent of three offshore oil fields in Abu Dhabi in 2018.

Figures for the oil, petroleum products, and natural gas pipelines industry provide a further unequivocal signal, albeit indirectly. In the 2018–2022 period, the United States and Russia are the biggest spenders. The first is rolling out an investment of $88.4 billion, the second $78.8 billion (GlobalData 2018).

The figures provide indisputable morally relevant evidence that the oil industry did not effect change—and does not appear to be planning to either—on its fossil fuel–centered behavior. In 2018 the major oil companies, both IOCs and NOCs, invested a sum equal to $50 billion in projects that are largely incompatible with the 1.5°C goal of the Paris Agreement (CTI 2019b), while the industry as a whole invested only 1.3 percent of its 2018 capex in low-carbon energy production (CDP 2018) and 0.8 percent in 2019 (IEA 2020c). These are hardly leaps and bounds toward the much-vaunted greening of their future output. Overall, the oil majors are projected to spend $785 billion on new oil and gas fields between 2020 and 2029. All capex in new fields is likewise irreconcilable with any climate goal (Global Witness 2019).

There is something else indicative in oil and gas companies' behavior: they have long been aware that climate impacts could endanger their business. And in their long-term business and operation planning they have prepared to brace for such a reality by taking into account potential climate impacts. For instance, back in 1989 Shell changed the engineering design of its offshore oil drilling platforms to account for sea level rise, and in 1995 Imperial Oil, a Canadian Exxon subsidiary, started considering the impacts of climate change in the Arctic in its planning strategies (CIEL 2017). Additionally, the industry is actively preparing for an impending climate crisis through the deployment of adaptation strategies for climate risk management. The most important include project design and location planning, emergency/crisis planning, risk management systems, and water management (IPIECA 2013).

FACT C: CAPACITY

Less carbon-intensive alternatives are available and have been for some time. Studies show that some major IOCs have actually had the capacity and the opportunity for more than forty years to reduce the harmful effects of their activities by modifying their business models (Frumhoff, Heede, and Oreskes 2015).

However, the oil industry, by and large, did not take any significant measures to reduce the harmfulness of its products, nor did it engage in policy redesign. Rather, as shown in the ensuing section, by denying climate change, the oil industry actively hindered such initiatives. The largest IOCs, however, have long been carrying out research to discover technologies to mitigate climate change. In particular, since the 1950s they have been studying and patenting technology to remove CO_2 from waste streams and carrying out tests on low-emission vehicles, fuel cells, and solar panels (CIEL 2017, 19–21).

CO_2 removal technologies have long held the attention of the oil majors, fully aware as they are of their potential for addressing climate change. Both Exxon and Shell had several patents for capturing and storing CO_2. But an initial obstacle slowed down and eventually brought to a halt the full development and industrialization of these technologies: "Removal of only 50 percent of the CO_2 from stack gases would double the cost of power generation" (DeMelle and Grandia 2016).

IOCs also invested heavily in fuel cells, which use the chemical energy of hydrogen or another fuel to efficiently produce electricity. Exxon and Shell, their attention piqued by a growing interest in clean and electric vehicle technologies, led this research in the early 1960s. The crisis that hit the oil industry in the early 1970s helped spur research into solar technologies: the 1974 US Solar Energy Research, Development and Demonstration Act distributed $6 billion in federal research subsidies in this area. American IOCs ended up netting the lion's share of those subsidies, either through getting in on solar energy research and development or by buying smaller preexisting solar energy companies. By the end of the 1980s the US oil industry owned or controlled the largest share of solar panel production in its homeland, maintaining its prominence in this technology well into the 2000s.

The largest oil and gas companies held the technical capacity for clean energy well within their grasp; proof positive is the fact that they were the proud holders of several early patents on a number of technologies that would have helped reduce their carbon output. If these technologies had been developed and deployed, Big Oil could have had a major impact in reducing carbon emissions and accelerating the shift toward becoming

Big Green Energy long before there was a moral imperative to do so. But the call of the shareholder's wallet seems to have more reach than the call of safeguarding Earth, so the prospects of the higher costs of carbon-saving technologies, at least initially, slashing the oil industry's profits meant that any plans to go down this path were shelved (CIEL 2017, 22).

It is worth reiterating that an alternative vision did actually exist, at least in the designs for the future expressed by some of the industry's more enlightened executives. For example, as already mentioned in chapter 1, in a 1997 speech given at Stanford University, the CEO of BP, John Browne, acknowledged the scientific consensus on anthropogenic climate change presented by the 1995 second IPCC report (IPCC 1995) as well as BP's responsibility and duty to take action as a consequence. Browne even remarked upon the potential of solar energy and affirmed BP's intention to invest in it, with projections that $1 billion in sales would be reached within the subsequent decade (Browne 1997). Browne's speech was widely lauded and raised hopes of a turning point for the industry. Observers saw it as being as revolutionary as the tobacco industry's acknowledgment of the correlation between smoking and cancer and heart disease. Unfortunately, a molehill was made out of a mountain. His words generated some fanfare in the media, with praise being heaped on Browne for his visionary stance by representatives of the environmental world, such as the Sierra Club and the California Environmental Protection Agency; other oil companies (e.g., Shell, Chevron) openly pledged to move in the direction outlined by BP and to put an end to climate denial. In the end, however, it turned out to be much ado about nothing.

It is clearly impossible to predict what might have been, precisely which different path might have been trodden, or indeed how much climate harm could have been averted had the oil industry fully developed and implemented the cleaner technologies it possessed. What it is reasonably sure, however, is that several decades ago major IOCs already had both the capacity and the opportunity to begin the process of decarbonizing their business and to markedly influence the behavior of the industry as a whole, with the obvious domino effect on the rest of the global socioeconomic systems. But they allowed this opportunity to shed the guise of Big Oil and become Big Green slip through their fingers, carrying on with business as usual.

FACT D: DENIAL

Science is complex; every caution is applied before affirming its veracity. Indeed, the intricacies and nuances of science may be hard for policy makers to grasp, let alone the general public. And given its abstractedness and remoteness in time and space, climate change science often proves to be a thornier issue than more linear disciplines. Scientific results are an ever-mutating entity, and therefore uncertainty will be their constant companion. Unfortunately, the public, by and large, has little tolerance for uncertainty, fueled to no small degree by a compliant media, lobbyists, and partisan lawmakers.

James Hansen, the former head of NASA's Goddard Institute for Space Studies, has frequently pointed out that climate scientists have somehow failed in mobilizing the public—let alone in engaging politicians—to act on climate change projections that they have been making since the 1980s; the doubt that remains rife in the minds of citizens does not disprove his theory. Nor have scientists succeeded in lessening the general aversion to uncertainty in relation to climate change, as elucidated by Nathaniel Rich's lengthy article to which the *New York Times Magazine* dedicated an entire issue in August 2018 (Rich 2018). Climate deniers and antienvironmental lobbyists have taken full advantage of this seeming lack of agreement on the basic elements of climate change theory, exploiting it to either deny it is happening or to repudiate the almost unanimous scientific consensus that climate change is caused by anthropogenic activity (Cook et al. 2013; Cook et al. 2016; Santer et al. 2019; Myers et al. 2021).

Much ink has already been spilled on the features and dynamics of oil companies' climate denial, which is profoundly wrong in moral terms.

Oreskes and Conway, in their 2011 *Merchants of Doubt: How a Handful of Scientists Obscured the Truth on Issues from Tobacco Smoke to Global Warming*, masterfully analyzed the oil industry's practice of investing heavily in climate change denial. While a brief reiteration of its main facets is called for—financing and orchestrating multiple initiatives for sowing doubt and misinformation about the existence and severity of climate change and the role that anthropogenic carbon emissions played in it, its science, and the motives of those who study climate change and communicate their findings—it would be more pertinent in this context to aim the spotlight on one of the main objectives of the industry's denial campaign: impeding and/or slowing action to address climate change.

Starting from the 1980s, the API prepared the ground by disseminating false and misleading information about climate change (Franta 2021). On this basis, leading IOCs actively opposed and, in many cases, successfully prevented policies on emissions reduction. To this end, since the early 1990s major IOCs have been deftly orchestrating a campaign of deception and disinformation—still enduring—with the primary objective of manipulating and steering public decisional processes to rein in fossil fuels. This campaign was used with great effectiveness to block regulations against fossil fuels and to refute the liability of the oil industry, mirroring what happened a few decades earlier within the tobacco industry (Oreskes and Conway 2011).

The Union of Concerned Scientists' (UCS) "*Disinformation Playbook*" is a practical point of reference (UCS 2018a). It lays out the IOCs' strategy for disproving climate science in order to oppose climate initiatives, articulated, it says, in five "*plays*" that read like the plotline from a major heist movie but help accumulate riches even Hollywood screenwriters would be hard-pressed to envisage (UCS 2018a).

1. "*The Fake*: Conduct counterfeit science and try to pass it off as legitimate research." Exxon, for instance, funded external scientists to publish mediocre research results contradicting the original findings of its own scientists, who all agreed on the relations between fossil fuel combustion and climate change and on its threat (Nuccitelli 2015).

2. "*The Blitz*: Harass scientists who speak out with results or views inconvenient for the oil industry." For instance, conservative *free-market* think tanks funded by the oil industry have been accused of being behind 2009's *Climategate* smear campaign and the 2010 attack on climate scientist Michael Mann (Deaton 2017).

3. "*The Diversion*: Manufacture uncertainty about science where little or none exists." Oreskes and Conway (2011) dubbed oil and tobacco companies *"merchants of doubt."* As a now infamous tobacco industry memo stated, "Doubt is our product, since it is the best means of competing with the *body of fact* that exists in the minds of the general public" (University of California San Francisco Library 1969). Unable to just conjure doubt from out of a hat, major IOCs rallied around to host initiatives discrediting the science and disseminating misinformation: real science was scoffed at, dismissed as mere junk, while misrepresentations were offered in its place. IOCs' pseudoexperts' favored *modus operandi* was to herald a (nonexisting) division in climate science to acquiescent journalists and politicians, happy to pass on the *news* to already confused laypersons (Ley 2018). Merchants of doubt but also masters of gaslighting.

4. "*The Screen*: Buy credibility through alliances with academia or professional societies." Generally speaking, through its generous donations, the fossil fuel industry seems to have whipped into line great swaths of US academic work on climate policy and energy. Exxon, for instance, has funded and still funds established research institutions, such as Columbia University, to investigate science, policies, and technologies to address climate change (Jerving et al. 2015); the Texan giant has for years also sponsored the American Geophysical Society annual meeting (UCS 2016). The API has partnered with African American and Hispanic business groups to publish op-eds in local newspapers to build support for offshore drilling by emphasizing its benefits, especially in terms of job creation (Volcovici 2018).

5. "*The Fix*: Manipulate government officials or processes to inappropriately influence policy." IOCs have long lobbied against climate policy

and regulations in the United States to great effect (Brulle 2018; Vard 2018), also having significant international repercussions. Consider, for instance, Exxon's successful efforts against US ratification of the Kyoto Protocol (Supran and Oreskes 2017) and in 2014 how the Western States Petroleum Association—the top lobbyist for the oil industry in the western United States, which included BP, ExxonMobil, Chevron, and Shell among its members—used fake consumer groups with innocuous names such *California Driver's Alliance and Washington Consumers for Sound Fuel Policy* as *astroturf* front groups, part of a campaign to create an illusion of widespread grassroots support against climate regulation (CIEL 2017). In the United States, major IOCs have significant influence on the Republican Party: their grip over climate and energy policy is very strong.

A sixth play could be added: *Passing the buck*. Exploiting the entrenched mindset to deflect blame, adopted by oil majors, consists of framing the question of climate change as one of individual consumption-based responsibility, thus preventing the general public from understanding that the climate crisis is a structural problem largely driven by the oil industry's denial, misinformation, lobbying, and disablement of climate policy and legislation. In this way, oil and gas companies have been able to obfuscate their responsibility for climate change and to present themselves as suppliers, merely meeting the existing demand, rather than as the major underlying cause of the problem.

The ultimate objective of these plays was to oppose climate action and they were successful since they contributed to paralyze climate policy. At the international level, for instance, the Global Climate Coalition, a fossil fuel–backed lobby group active in the mid-1990s and early 2000s, used all of the abovementioned plays to manipulate the IPCC, the United Nations' official scientific advisory body on climate science (Hope 2019a). Some IOCs—especially Shell—actively tried to obstruct international climate negotiations thanks to privileged access to the annual United Nations Framework Convention on Climate Change meetings through trade associations (Hope 2018).

A second worryingly effective consequence of some IOCs' funding of climate denial is crucially the increasing polarization of the climate discourse generated by the complex relationship between politics, science, and climate scientists (Stern et al. 2016; Hansson 2018). Climate denial is far from over; indeed, in its new form it is thriving. From 1986 to 2015, the five biggest fossil fuel corporations in America spent $3.6 billion on advertisements basically claiming that fossil fuels are virtuous and necessary and that the oil industry is actually addressing climate change (Atkin 2019). In recent years Europe too is experiencing a proliferation of climate-contrarian think tanks that use largely the same arguments and activities described above traditionally employed by the US climate change countermovement (Almiron et al. 2020).

This snowballing of the denial machine, still largely propelled by major IOCs (Frumhoff et al. 2015), could further polarize the climate discourse by influencing the contents of denial themes, widening their scope and their prevalence over time (Farrell 2016; Cann and Raymond 2018).

Ironically, the consensus among scientists with regard to climate change has engendered skepticism among denialists, now almost bordering on the intolerant. This skepticism can, in fact, be more properly defined as climate *cynicism*: doubts about the evidence produced by climate science have been skillfully replaced by *ad hominem* doubts about the people who study climate change and communicate their findings and their motives. Cynicism is fueled by the ease with which such doubts are sowed in a fertile terrain that requires the right balance of time, money, and political context. ExxonMobil alone has *invested* breathtaking sums—more than $240 million—to do so in the last two decades.

At the same time, a political context able to politicize the scientific orthodoxy—not simply to dispute it—was carefully curated with the vast cash injections from some IOCs (Thomas 2017). The new breed of denialists have been able to portray climate change as an issue created by climate scientists *for* climate scientists, as a fabrication to keep alive an obscure yet substantial techno-scientific elite, as well as for the good of the well-known pro–big-government/higher tax rate environmentalists

(Hoffarth and Hodson 2016). IOCs' money and effort can be deemed a mission accomplished: climate change has become a question of political tribalism and turned into an emblem of the partisan divide. Indeed, this current polarized political mindset—accompanied in no small way by a fractured media—works fully in Big Oil's favor. The denial machine only needs to sow the seed of doubt and throw its fossil fuel–derived money behind the most accommodating political decision makers and then can sit back and watch public opinion be swayed, thanks to the *pro hominem* fallacy, a kind of *honor by association* perpetuated by the tribal loyalty that prevails over reason.

Unfortunately, no game changer seems to be in the cards. Major IOCs (BP, Chevron, ExxonMobil, Shell, and TotalEnergies) invested over $1 billion of shareholder money in the three years following the 2015 Paris Agreement on misleading climate-related branding and lobbying (InfluenceMap 2019). In particular, they spent €251 ($283.2) million lobbying the European Union between 2010 and 2019 alone. Hundreds of American and European—largely UK-based—individuals and institutions involved in climate denial signed a letter in late 2019 to leaders of the European Union and the United Nations arguing that there is no climate emergency and therefore no need to set net-zero emissions targets. In recent years FTI Consulting, one of the most notorious oil industry public relations firms, has had a prominent role in influencing campaigns to support fossil fuels. For instance, this firm helped design, staff, and run organizations and websites funded by oil companies that can appear to represent grassroots support for fossil fuel initiatives (Tabuchi 2020). The same FTI Consulting is behind the prohydrogen push in Europe: it presents hydrogen as a clean fuel, while in truth it is still mostly produced through Big Oil's methane (Mikulka 2020).

Additionally, the oil industry has a new denial tool since it has started using *discourses of climate delay* that, by downplaying the urgency of the climate crisis and overstating the industry's progress toward addressing climate change, justify inaction or inadequate efforts (Lamb et al. 2020).

In brief, via an intensive, systematic, and sophisticated denial campaign, major IOCs have successfully opposed any political efforts to move socio-economic systems away from fossil fuels, thereby inducing decision makers to commit a morally relevant omission that has seriously aggravated the negative repercussions of the climate crisis on a global scale.

FACT E: ENRICHMENT

An indisputable truth overshadows any debate regarding the oil industry: the majors have made substantial profits that have seen them acquiring extraordinary wealth through their fossil fuel–related activities, as testified by the *Polluter Elite Database* (Kenner 2019), which reports the shares detained by large multinational oil, gas, and coal companies' executives and directors as well as the values of their shares and their personal emissions related to the ownership of such shares.

Few would dispute that this alone is either morally wrong or, per se, related to harm. However, fact E, enrichment, is still morally relevant, since it strengthens and better exemplifies oil companies' moral responsibility for climate change.

As mentioned above, in this context to see why the wealth accumulated by the oil industry sets a different and complementary moral scenario that more effectively shapes its moral responsibility for climate change, a brief look at the moral principles that justify oil companies' rectificatory actions is required.

In this regard, climate ethics literature (e.g., Caney 2005; Shue 2015) usually makes reference to two backward-looking principles, the *polluter pays principle* (PPP) and the *beneficiary pays principle* (BPP), and one forward-looking principle, the *ability to pay principle* (APP). The PPP allocates financial and other burdens associated with rectificatory actions in proportion to past contributions that agents have made to the overall level of harm. On the other hand, the BPP holds that this allocation should be calculated based on the benefits that agents have derived from activities generating harm. Finally the APP posits that the quota of burdens should be proportional to the agents' relative capacity to withstand the encumbrance.

While the morally relevant facts A, B, C, and D described in the previous sections are all related to harm and therefore refer mostly to the PPP, fact E, which is not related to harm, refers to the BPP and the APP. In other words, the inclusion of the *wealth component* intrinsic to fact E reinforces the justifications for the oil industry's moral responsibility, especially in view of a consequent *duty of reparation* that should take the form of a disbursement of funds. Given the need to involve the oil industry in climate policy and governance, the more its responsibility is articulated, the more cogent it is likely to be. In practical terms, the question of the oil industry's wealth should be quantitatively addressed referencing companies' profitability trends. For instance, BP, Chevron, Exxon, and Shell in the thirty years from 1990 to 2019 accumulated $1,991 trillion—BP $332 billion, Chevron $360 billion, Exxon $775 billion, and Shell $524 billion—in profits (Taylor and Ambrose 2020). However, profitability in the oil industry depends on a myriad of contingent economic, social, political, institutional, and environmental factors as well as on deliberate internal financial and fiscal choices; therefore, profits vary greatly over the years, and profitable and nonprofitable periods tend to span several years.

All industries experience fluctuations in both long-term and short-term periods of profit and loss, and the oil industry is no exception. For instance, the first quarter of 2018 was the most profitable in years for IOCs, mainly due to a marked increase in oil prices and to the industry's success in cutting costs. In 2018, Saudi Aramco posted total net profits of an astounding $111.1 billion. Here are some other key figures of major IOCs' 2018 first-quarter profits (Cunningham 2018):

- BP's profits soared by 71 percent to $2.4 billion, compared to $1.4 billion a year earlier;
- Chevron's profits went up to $3.6 billion, an increase of 36 percent compared to 2017;

- ExxonMobil saw its profits rise by 16 percent to $4.7 billion compared to 2017; and
- Shell's profits surged to $5.32 billion, 42 percent more than the same trimester in 2017.

The oil industry's wealth can probably be further grasped, albeit rather allusively, by examining industry-linked individuals who have accumulated extraordinary wealth through fossil fuels: the *oil billionaires*, usually with close ties to government-run NOCs.

Russian president Vladimir Putin has a fortune estimated at somewhere between $40 billion and $200 billion, and most of it can be traced to his stakes in the oil sector (Harding 2007; Calcuttawala 2017). He is rumored to own 37 percent of Surgutneftegas (a Russian oil and gas company with reserves in western Siberia created by merging several previously state-owned companies) and 4.5 percent of Gazprom. A fortune with similar origins is detained by Azerbaijan's president Ilham Aliyev, whose AtaHolding, according to the Panama Papers, held over $490 million in assets, mostly in the oil and gas sector (Fitzgibbon, Patrucić, and Rey 2016) and by the former Kazakh president Nursultan Nazarbayev. Isabel Dos Santos, daughter of the former president of Angola, a destitute country with massive oil wealth, chairwoman for a year of Sonangol, Angola's NOC, is worth $4.3 billion. Conservative estimates from various sources of the undisclosed wealth of Sultan Hassanal Bolkiah of Brunei—the third-largest oil producer in Southeast Asia—puts his assets somewhere between $20 billion and $40 billion, most of which has been accumulated by the exploitation of the country's huge reserves of oil and gas.

The morally relevant facts from A to E, however, remain largely obfuscated by the almost endless number of factoids—in the original sense of the Norman Mailer–coined neologism meaning something that sounds credible and is assumed to be true by a significant number of people and yet is not true—that the oil industry and, more broadly, those who for diverse reasons contest the realities of climate change have disseminated over the last decades. At the same time, it is dispiriting to fully grasp the power of factoids: despite essential public interest in health and safety and the well-being of humans and the planet being at stake, the oil industry could and does defend and advance its vested interests by denying science, browbeating scientists, and subjugating politics with their shell game—no pun intended—of deftly mixing plausible factoids in with the indisputable facts.

II   BIG OIL'S RESPONSIBILITY AND DUTIES

Much as the planets in our solar system revolve around the sun—the primary source of its energy—our socioeconomic system orbits around the industry that feeds its current carbon-intensive needs: Big Oil. But this kingpin of industry, as emphasized in part I, is also a fulcrum of the climate crisis. A useful perspective for addressing this role and the possible entry points to change it is to investigate the behavior of oil and gas companies through a moral prism focused on the industry's responsibility and duties. Before proceeding, this claim and its centrality to the argument of the book need to be clarified.

The rationale of laying out a moral framework for attributing responsibility and ensuing duties to Big Oil is that this analysis can provide the basis for justifying and reinforcing the claims of justice for civil society's grievances against fossil fuel companies for climate change, rendering them more politically acceptable and thereby effective.

Indeed, climate activists and, more broadly, civil society have started to train their focus on fossil fuel companies and projects. They make two standout demands of Big Oil: to rectify the harm already done and to discontinue fossil fuel production in order to avoid further future harm.[1]

This is also consistent with the human right to a healthy environment as a United Nations Panel of Experts report suggests (UN 2018). Additionally, a report by the United Nations special rapporteur on extreme poverty and human rights highlights the role of fossil fuel companies in impeding action on climate change as a patent violation of human rights (HRC 2019).

The core claim of climate justice movements is, in fact, that richer agents, including corporations, repay their climate debt, divided into an *impacts* debt and an *emissions* debt. The ultimate objectives of this request—which in many respects are consistent with the United Nations Framework Convention on Climate Change (UNFCCC) core ethical ambitions—are to take democratic control over the economy, govern climate change in a participatory way, and lessen the injustices involved. The impacts debt implies, by and large, a rectification of the harm brought about by climate change, while the emissions debt requires action to reduce carbon emissions overall and the associated future harm, possibly in conjunction with some form of historical contribution to the problem as demanded, for instance, by the Lofoten Declaration, signed by 530 organizations from seventy-six countries representing business and industry, civil society, universities, research organizations, foundations, cities, and religious institutions.

In a different and complementary perspective, the normative theorization of this part of the book aims at providing the moral ground for climate action to have a positive influence in the real world. Without entering into the thorny methodological debate about the capacity of climate ethics approaches to achieve this objective, suffice it to say that the moral framework developed here refers to the so-called *engaged methods* of climate ethics (Green and Brandstedt 2021). As such, it is equipped to contribute to real-world climate action since it involves clear identification and substantial engagement with the agents of change, in this case the first-order agent Big Oil and those who in chapter 4 are dubbed indirect (or second-order) agents (e.g., political authorities at various levels, members of the civil society, economic actors, epistemic communities). Basing the normative theorization on the distinctive, structural role of first-order and indirect agents with respect to the climate crisis suggests possibilities for normative-prescriptive work that, by lessening the indeterminacy of the moral framework proposed, can more effectively mobilize action at different levels. In fact, a moral framework of this type considers the political possibilities of first-order and indirect agents' actions of change in a situated and detailed way, thereby closing the gap between normative theorization and its real-world positive influence.

In light of these considerations, a framework of Big Oil's responsibility and duties as laid out in the following chapters can serve as a useful foundation for facilitating the consolidation of emerging anti–fossil fuel social/moral norms condemning the industry for its deliberate engagement with such a harmful product. Part III of the book argues that civil society and other stakeholders can better engage and disrupt in a democratic and nonviolent way oil and gas companies' power to favor the introduction of binding provisions compelling them to address past harm caused by their activities and products and to steer their future behavior toward less harmful business models, as dictated, respectively, by the duties of reparation and decarbonization.

By no means does this book imagine that laying out a moral framework for oil and gas companies in climate change will become a sort of self-fulfilling prophecy whereby the industry will simply comply with recommendations in light of the *authoritativeness* of the ideas of morality, harm, responsibility, and duties laid out. This book acknowledges that the framework of morality developed is not sufficient in itself to motivate oil and gas companies to take action. But in the highly politicized world of the oil complex, one of the book's main goals is to lay the groundwork for investigating and justifying modification of the social, economic, political, and legal context the oil industry operates in necessary to lessen its harmful behavior.

In a nutshell, the moral framework laid out in this second part of the book provides justification for assigning oil and gas companies' responsibility for and subsequent duties to engage with the climate crisis, with part III establishing the external and internal motivations that could generate impetus for them to actually address the harm—both past and future—associated to their activity.

This chapter explores the main implications of climate ethics for Big Oil, situating and delineating its duties beyond the traditional adaptation/ mitigation dichotomy of climate studies and emphasizing the unique agency of the industry. To this end, climate change must first be framed as an ethical issue, with a series of controversial moral features of the oil industry being examined.

CLIMATE ETHICS: MAIN ISSUES FOR BIG OIL

That climate change is an ethical issue should strike few as surprising. Philosophers as well as politicians, climate activists, and religious leaders, among others, have long highlighted the numerous ethical challenges that are inseparable from considerations on the causes, consequences, and potential human responses to global anthropogenic climate change. Indeed, the very quintessence of this chapter is embodied in two oft-quoted claims that still hold true to this day.

Al Gore, who was jointly awarded the 2007 Nobel Peace Prize with the International Panel on Climate Change for their efforts against climate change, affirmed that climate change "is not a political issue. This is a moral issue, one that affects the survival of human civilization" (Gore 2007). Similarly, James Hansen wrote that "the predominant moral issue of the 21st century, almost surely, will be climate change, comparable to Nazism faced by Churchill in the 20th century and slavery faced by Lincoln in the 19th century" (Hansen 2011).

Climate change has long been acknowledged as fundamentally an ethical issue that threatens our lives and our world (Gardiner 2004). However navigating a climate-shaken world in a morally sound way raises a myriad of crucial matters. Despite the growing recognition that urgent social-ecological reconfigurations will need to be deployed to sustainably address the uncertainties of the *Anthropocene*—the current epoch, characterized by significant human impact on Earth's geology and ecosystems—explicit and agreed moral guidance to tackle the climate crisis is still sorely lacking. It may not currently be possible to know exactly what morally grounded actions are needed to avert the crisis, but it may be possible to outline, despite a landscape of high uncertainty—or even indefiniteness—the basic moral constitution to address some of the major threats to humanity and nature if the present unsustainable trends continue.

As societies move along their current trajectories, comprehension of the positive or negative consequences of a particular path may also influence present decisions about whether or not to tread it. However, this adaptive attitude is only possible if there is an agreed-upon moral consensus about the direction to be taken, a political choice that inevitably is morally connoted and truly of a normative nature. A lack of moral guidance could, in fact, beget a paralysis of policy and governance and worsen moral corruption in engaging sustainably with the climate crisis.

In the moral and cultural milieu of the Western philosophical tradition—which although clearly not superior to other perspectives is widely acknowledged around the globe and has largely contributed to the formation of existing, albeit weak, global governance institutions, including those addressing climate change—a convenient starting point for the construction and vindication of climate ethics is the consideration of the liberal account of justice. The reference is to modern liberalism, which is based on equality, freedom, redistribution, inclusion, and care. Modern liberalism gives equal or impartial consideration to the interests of all and displays a general concern for the least well-off agents, who should be given the opportunities, means, and choices to live a dignified life, the improvement of which is the most ethically important objective. This concept is the nerve center of liberalism and is possibly the central tenet of the dominant ethics of climate change, including the moral issues specifically related to Big Oil.

Climate ethics, which generally runs on considerations of distributive justice and problems of burden-sharing (Caney 2014; Vanderheiden 2016), has addressed different intertwined issues through the adopted lens of liberalism: Whose responsibility is it? What is owed to whom?

What does the current generation owe to future generations? Do richer agents owe poorer ones due to their emissions and harm as well as to their greater capacity? How should the cost of confronting and/or preparing for climate impacts be shared? What principles justify burden-sharing regimes? How can policy makers make morally sound decisions, given the great uncertainty of climate change?

Most of the topics above have entered the mainstream climate debate in earnest; at the same time, climate ethics faces further novel challenges that affect foreseeable evolutions of a world threatened by a climate emergency. In this regard, engaged climate ethics should focus on topics that cross disciplinary boundaries, as these coincide with the most pressing issues that policy makers, negotiators, regulators, businesses, nonprofit organizations, climate communicators, and others involved in responding to climate change will face in the coming years. The related focal points concern, piecemeal, the responsibility of agents, the scope of ethical considerations, the possibility of collective agency, and the operational potential of climate ethics. Most of the relevant moral challenges concerning Big Oil in relation to climate change are included in or touched upon by these issues.

Responsibility

In the labyrinthine path to addressing climate change, one of the thorniest issues—and possibly the major culprit behind the long deadlock of international negotiations—is the assignment of responsibility in terms that are simultaneously morally, legally, economically, politically, socially, and psychologically acceptable to policy makers and the broader public. The UNFCCC has relied on the principle of *common but differentiated responsibilities* to outline, albeit vaguely, how the burden of tackling climate change is to be apportioned. Although this language can be appropriate in diplomatic contexts, the implication that all agents have responsibilities and duties has produced a *you-first* attitude among policy makers and negotiators. At any rate, responsibility is one of the most difficult and confusing terms of moral and political philosophy, even more so when it is applied to an unfamiliar and cognitively complex moral context such as climate change (Miller 2007, 2008).

Some authors use *responsibility* and *duty* interchangeably. As anticipated in chapter 1, this book instead distinguishes the two concepts, adopting the view that responsibility is the *condition* of being responsible according to principles of justice as well as the obligation to take action. On the other hand, a duty is a standard of moral behavior dictated by responsibility and involves a *practical commitment*—such as those outlined in chapter 6 and operationalized and implemented in chapters 9 and 10 of part III—to either do or refrain from doing certain things.

In broad moral terms, responsibility relates to an agent's conduct and intentions. Numerous are the notions and perspectives of responsibility pertinent to Big Oil, analyzed in chapter 4, which directly addresses this problematic. Suffice it to say here that a fundamental distinction to make in relation to Big Oil's responsibility is between its *causal* and *moral* understandings. Causal responsibility is basically intended as *causal contribution* and is less stringent than moral responsibility, which is based on the appraisal of agents' intentions, assessing their voluntariness, control, and knowledge: in this view, to some extent, chapter 1 laid the foundations for the oil industry's causal responsibility, while chapter 2 investigated the facts rationalizing its moral responsibility.

In the face of such complexity, perhaps one of the most important notions to bear in mind when attempting to establish Big Oil's responsibility for the climate crisis is that there is no single, salient, widely agreed-upon notion, since its constructs are shaped and then used for particular purposes (Jamieson 2015). As a result, a great deal of plasticity exists in terms of how the issue can be construed by different individuals, groups, and organizations. In the same way, the notion of responsibility—developed and applied to the oil industry in the following chapters—is based on highly context-specific morally relevant facts and, by and large, should not be applied to other agents or settings.

Scope

One question above all provokes debate when discussing climate ethics: *Who counts?* And what that boils down to is *which agents should be at the center of the climate debate?* Beyond the current state-centric international perspective, which considers states as the only agents, there is a bubbling cauldron in which a number of alternative nonstate agents get thrown into the mix.

A prime example of this is the focus on individual agency in terms of both reducing ones' own emissions and advocating for larger-scale change. Although this perspective has gained some traction in recent years especially among environmentalists, there are also important questions about how such rationalizations actually influence engagement with the issue among individuals and communities. Even among the highest individual emitters (e.g., upper- and middle-class Westerners), it is unclear how far people feel responsible for their emissions. Intersecting with these psychological questions are ongoing normative ethical questions about how much responsibility individuals actually do have for the harm caused by their emissions (in the grand scheme of things minuscule: according to the International Energy Agency [IEA 2021] individual behavioral changes would only account for about 4 percent of the reductions necessary to achieve a net zero target by 2050), as well as positive moral questions regarding individual responsibility, given political and economic constraints on action as the sixth play of oil industry climate denial described in chapter 2 evinces.[2]

For example, works on the contribution of major carbon producers to global cumulative emissions (Heede 2014) and to surface temperature and sea level (Ekwurzel et al. 2017) and ocean acidification (Licker et al 2019), mentioned in the previous chapters, make it difficult to argue that individuals are the sole or even main agents responsible for causing or addressing climate change. It is therefore necessary for climate ethics to better explore forms of collective responsibility that do not negate individual responsibility but instead integrate the two perspectives in ways that allow the climate regime to deal with the individual and the aggregate level in one fell swoop, perhaps with particular attention to newly considered agents of justice such as the corporate entities under discussion that have so far remained largely overlooked.

Collective Agency

In light of the above, the objective of analyzing the responsibility of Big Oil in climate change requires reflection on collective agency and collective responsibility as central constructs for a broader and more effective moral analysis of the climate crisis. Collective responsibility is a controversial concept, since standard ethical views assume that individual human beings are the ultimate moral agents and that groups of individuals—communities, corporations, states, nations, and international institutions—are only indirect moral agents.

At any rate, there are different theoretical perspectives that justify the collective responsibility of groups. In this context, a useful one is French's (1984), which attributes collective responsibility to *conglomerate collectivities*, that is, organizations of individuals whose identity is not limited to the sum of the identities of the persons in the organization. Conglomerate collectivities have the following four features: an identity that is greater than the sum of the identities of their members, decision-making structures that enable the input of members' decisions to be translated into collective decisions as outputs, consistency over time, and self-conception as a unit. This book categorizes oil and gas companies as conglomerate collectivities because they meet all four of the above criteria. Additionally, the book also assumes that oil and gas companies are fit to shoulder responsibility, as they satisfy the requirements that their choices are value-relevant, they are able to make the appropriate value judgment, and they are value-sensitive because they have the ability to act on the judgment made (Pettit 2007).

This book therefore holds that the fossil fuel companies are to be considered fully fledged collective agents of climate ethics. The acknowledgment of collective agency and responsibility of oil and gas companies in climate change does not, however, dispense with individual responsibility, which operates at a different analytical and practical level, not considered in this book. In other words, although the oil industry is collectively responsible for climate harm and its rectification and removal, other agents also undertake behavior that adds to carbon emissions. It is worth reiterating that by no means does this argument imply that Big Oil should become the sole agent responsible and duty-bound for addressing climate change or that oil and gas corporations are the most important players. States, consumers, civil society, different industries, and other stakeholders all have a role in climate change and consequent responsibility and duties to do their part to resolve it.

The Feasibility Issue

As a closing note, this section will briefly—and indeed only partially—touch on the critical point of moving climate ethics into practice. This issue centers in large part on the relevance of the questions being asked by negotiators, policy makers, citizens, institutions, and society at large; perhaps one of the aims of this book is to represent a means to an end to achieve this goal.

The inclusion of climate ethics in decision making must therefore be supported by meaningful reflection in light of current institutional and political understandings and constraints. To be beneficial, such work must provide accessible, stable, and actionable insight into the problems faced by decision makers rather than providing answers to questions no one is asking outside the ivory tower. At the same time, climate ethics has the potential to help shape these very questions being put forward *out there* and, in so doing, helps make climate change decision making more inclusive and ethically grounded at various stages of the process.

So in this spirit, this book lays out the duties of reparation and decarbonization deriving from oil companies' responsibility for the climate crisis. Said duties are defined as moral provisions, with immediate practical relevance, that splice together the normative perspective of responsibility to the positive perspective of climate policy and governance.

ADAPTATION, MITIGATION, AND BIG OIL'S DUTIES

Climate change is having an array of negative impacts on our planet's natural and socioeconomic systems, directly or indirectly harmful to all of humankind and potentially catastrophic for many of the most vulnerable people in the world.

The most prominent of these regionally differentiated human-threatening impacts include greater water stress leading to reduced crop yields; rising sea levels; increased inland floods and coastal flooding and erosion; reductions in the thickness and extent of glaciers, ice sheets, and sea ice; exposure to new health risks; rises in the frequency and severity of extreme climatic events; and increased conflicts over the control of scarcer resources, migrations, and state failures and the resulting risks. For humanity, such diverse impacts threaten food security globally and regionally; increase risks from foodborne as well as animal-borne, water-borne, and vector-borne diseases; intensify the displacement of people due to migration; increase risks of violent conflicts and wars; reduce economic growth and poverty eradication; and create new poverty traps (IPCC 2014; National Intelligence Council 2021).

As underlined in part I, oil and gas companies, through their fossil fuel–related activities, have undoubtedly contributed to these impacts and therefore have concurred to harm the planet and humanity. Averting the disastrous implications of the current trend of climate change requires avoiding/preventing harm, as the objective of the UNFCCC implicitly acknowledges. The requirement *not* to do harm is the fundamental component of climate ethics (Shue 2011, 2017), the main imperative of which is to prevent people from suffering climate-related harm (Vanderheiden 2011).

As indicated in chapter 1, Big Oil concurred in harming the planet and humanity through the emissions caused by its activities. At the same time, chapter 2 clarified that Big Oil erected obstacles to the acknowledgment of the harmfulness of its climate-related activities and of climate change itself, thus further enabling harm, as detailed in chapter 4.

However, the current increasingly sophisticated literature on the ethical implications of the climate crisis by and large does not seem to consider harm as its central moral tenet; indeed, it prefers to apply a resource-sharing perspective centered on the allocation of costs and benefits of actions related to climate change, largely independent from considerations of harm. This dominant perspective contends that climate change entails two moral commitments: first, to curb anthropogenic greenhouse gas emissions and augment carbon sinks in order to avert dangerous interference with the climate system, and second, to support and fund efforts aimed at preventing or coping with climate impacts These are known as the duty of mitigation and the duty of adaptation, respectively, and are both subject to intense debate in the burgeoning literature on the issue.

This book, however, argues that in the moral discourse on Big Oil in relation to climate change, both the duty of mitigation and the duty of adaptation are, so to speak, instrumental. In other words, they are a means for dealing comprehensively with the harm resulting from climate impacts, the ultimate end of the struggle against climate change. In fact, the only way to avoid/prevent harm associated with climate change requires both protecting nature from society (i.e., avoidance of harm) and society from nature (i.e., prevention of harm). In particular, both harm avoidance and long-term harm prevention depend almost exclusively on mitigation efforts, whereas short-term prevention largely depends on adaptation measures. Consequently, mitigation and adaptation commitments are two sides of the same coin, as they both ultimately address a single, fundamental moral issue, namely avoiding/preventing certain agents from harming other agents, the moral core of climate change.

Given the establishment of a connection between Big Oil's activities and its contribution to global cumulative carbon emissions leading to harmful climate impacts, an analysis to establish the duties that moral reasoning assigns to the industry can be framed in terms of responsibility. This provides the moral basis for adopting an integrative approach to the duty of reparation and decarbonization developed in the book. In essence, these duties are respectively none other than manifestations of the traditional duties of adaptation and mitigation when framed in terms of harm with specific regard to oil and gas companies. The duty of reparation encapsulates the requirement that Big Oil rectify the injustices resulting from the harm the industry has generated, while the duty of decarbonization entails an obligation by the industry to eliminate carbon emissions from its activities to prevent future harm.

Both duties will be examined—theoretically and empirically–in finer detail in the remainder of the book.

UNIQUE AGENCY OF BIG OIL

Part I shows that Big Oil has contributed to the climate crisis by causing, shaping, advancing, and defending the current unsustainable fossil fuel–dependent global economy. Through its informed and self-advantageous choice—backed in no small way by big dollars—to continue the exploration, production, refinement, and distribution of fossil fuels even after the harm associated to their combustion became undisputable, Big Oil has essentially imposed on the global socioeconomic system a carbon-intensive model of development rather than engaging in a concerted search for alternatives and phasing out fossil fuels, as warranted by the urgency of the climate crisis. In this light, it is morally unacceptable to equate Big Oil's responsibility with those of other agents. Climate governance should therefore reflect the unique agency of Big Oil, as it has played a very singular and significant role in the climate crisis and should contribute to tackling it accordingly.

Despite the oil and gas companies' substantial contribution to the problem, the wealth and benefits obtained through fossil fuel–related activities, their political influence, and the technical expertise that would grant them a relatively smooth transition to less carbon-intensive products, these companies currently have no special input in the climate governance and policy framework. Likewise for other corporate agents, they are only subject to the binding emission limits on their processes, imposed by national and subnational political authorities. At best, like other industries outside the carbon business, on a voluntary basis Big Oil agrees to disclose its carbon emissions and to integrate abatement strategies in its processes. Given the nature of its core business, though, this is far from enough.

Scientific knowledge and consensus about climate change have reached a stage whereby fossil fuels should be deemed a harmful product, the use of which is affecting health and lives and the well-being of the present and future generations of humans and nonhumans. There have been previous examples in history where the harmfulness of a product was confirmed by solid scientific evidence, provoking a reshaping of entire industries. Like companies trading in tobacco, asbestos, or lead-based paint, Big Oil should assume some form of responsibility and countenance any related duties stemming from its involvement in dealing in a harmful product.

Part I draws attention to Big Oil's specific contribution to climate change and offers a basis to build a normative case for the duties to address the problem. Including oil companies in climate governance and policy would extend the scope of the debate from an ossified and still prevalently state-centered UNFCCC perspective and would be consistent with the current increasing interplay between state and nonstate agents in climate governance, which disregards and challenges old geopolitical groupings and institutional structures.

Not all major oil and gas companies operate in wealthy states, as part I makes clear. This indicates how complex the structure of the current global economy is. A large amount of emissions have come from fossil fuels operated in less wealthy countries, such as Algeria, Brazil, India, Iran, Mexico, and Nigeria. Recognizing these countries' companies as important players in global climate change and holding them responsible for their fossil fuel–related activity would, among other things, help bridge a simplistic divide between *the rich* and *the poor* worlds. It could lead also to a fairer distribution of the burden of fighting climate change among state and nonstate agents around the world.

Introducing fossil fuel companies as moral agents in the context of climate change throws open the doors to a very useful avenue of inquiry in climate ethics, which could have major implications on climate governance and policy. For example, an alternative mode of recognizing the responsibility of the different agents in the global system could alter the approaches to rectification of harm caused and the related distribution of burdens and benefits, influence the patterns of well-being among agents, and change the flow of significant financial resources as well as other assets across peoples and generations.

When tourists approached Scotland's £75 ($102.3) million Glasgow Science Center on June 28, 2018, they were met with a curious sight: the futuristic titanium-clad roof was oozing thick, black, molasses-like goo—perhaps, they wondered, some publicity-seeking mass-scale scientific experiment was under way. Somewhat ironically, what was actually happening was that the building's weather-proof membrane was literally melting on the hottest day the city had ever seen, 31.9°C. Glasgow is on the 55th parallel north and sweltering days were once few and far between.

According to NASA, 2020 tied with 2016 as the hottest year on record globally, while 2019 was the third hottest. The extraordinary heat waves that hit Western Europe in the summer of 2019 set all-time temperature records in multiple places, with Paris, to name but one example, recording an all-time high of 42.6°C on July 23. Events such as the ones experienced in July 2019 would have been 1.5° to 3°C cooler in a climate unaltered by man-made carbon emissions and were made between three to ten times more likely by climate change (Vautard et al. 2019). In 2019, 2020, and 2021 Siberia, Brazil, Indonesia, Australia, and the western

United States experienced colossal wildfires exacerbated by hotter temperatures and drier weather conditions associated with climate change. The NOAA showed that in 2020 the country experienced 22 disasters, 262 dead, and $95 billion in damages caused by climate change (NOAA 2021). In the summer of 2021, the hottest on record both in Europe and the United States, a cascade of deadly weather events scourged the northern hemisphere: from the heat wave in Canada and the US Pacific Northwest, to the floods in China and Germany, and the fires in the Mediterranean, to name only the most extreme and devastating.

The knock-on effects of climate change are causing havoc to the planet: more heat, a wetter atmosphere, and more extreme rainfall are expected globally (Sun et al. 2021). The Atlantic meridional overturning circulation is weakening, and this could cause more frequent and intense heat waves in Europe, increase sea level rise in North America, and force fish to move north (Caesar et al. 2021). Ice is melting worldwide especially at Earth's poles, where climate is changing and temperatures are increasing at an extremely rapid pace (Landrum and Holland 2020). While in the 1990s our planet lost roughly 800 billion tonnes of ice each year, today, due to increasing air temperatures, that number has surged to around 1.2 trillion tonnes: altogether, 28 trillion tonnes of ice disappeared between 1994 and 2017 (Slater et al. 2021). Such massive melting has even caused a shift in Earth's axis of rotation since the 1990s (Deng et al. 2021).

Many species have been impacted by increasing temperatures; sea levels have been rising more quickly over the last century; precipitation has increased across the globe, on average; hurricanes and other storms are likely to become stronger; floods and droughts are more common; and less freshwater is available (IPCC 2014). In short, we could be dangerously close to crossing a planetary threshold beyond which it would be impossible to stabilize the climate to avoid a "*Hothouse Earth*" (Steffen et al. 2018), that tipping point at which heat becomes a killer for much of life on Earth.

These climate truth time bombs basically raise fundamental moral and pragmatic issues about the harm caused by Big Oil and about its responsibility and duties in the climate crisis. This book suggests that given the harm done, the oil industry—jointly with other agents—is responsible for climate change. This assumption, as anticipated, provides countless offshoot questions requiring an answer that this chapter tries to address. What is the moral status of the harm caused by Big Oil and its connections with the industry's responsibility? Why is Big Oil responsible for climate change impacts and its related harm? Based on its responsibility, how should Big Oil rectify the harm done and prevent further harm? Who ensures that Big Oil complies with its harm-related duties, and how?

HARM

Few would argue that climate change does not pose severe existential threats to people's fundamental rights and interests and to the planet they inhabit. Given this, it is useful to first clarify the moral status of the harm generated by Big Oil's climate-related activities before scrutinizing the connection between harm, responsibility, and duties to attempt to contextualize them in relation to the industry.

As said, the requirement to do no harm is a central tenet of morality and has shaped and guided societies for generations. The *do no harm* principle states that agents have negative duties, that they should eschew certain behaviors in order to prevent and avoid doing harm to others. By and large, harm arising from climate change is viewed as distant and abstract, so climate change is not perceived as a moral problem and does not prompt the usual urgent responses to moral challenges (Jamieson 2008). The human brain too is unprepared to respond to the challenges raised by the climate crisis since it has evolved to cope with more immediate threats that violate our moral sensibilities (Sacchi et al. 2014).

Climate-related harm associated with Big Oil has two distinctive features that may help delineate it and eradicate or at least lessen its intractability: (1) the perpetrators of harm—oil and gas companies, in this case—and their contributions are identified, and (2) attribution science—the burgeoning science of attributing weather events and its potential contribution to assessing loss and damage associated with climate impacts, further addressed in chapter 9—is making it increasingly possible to trace specific harm-generating climate impacts to oil companies.

Considering empirical evidence of the harm deriving from Big Oil's activities presented in part I, the industry is clearly in violation of the do no harm principle. In this light, a societal decree must establish the most appropriate forms of duties to be imposed based on the morally relevant facts proving violation of this principle.

But what are the relevant moral traits of the notion of harm employed in this book and its connection with responsibility? To this end, it is worth remembering that Big Oil concurred in doing harm to humanity and the planet through the greenhouse gas emissions caused by its processes and products and in enabling harm through the harm-related morally relevant facts A, B, C, and D—awareness, behavior, capacity, and denial—analyzed in chapter 2. The overall moral cogency of climate-related harm and its connection with responsibility originate from and can be dealt with within the doctrine of doing/allowing and enabling harm.

Reasons against doing harm (i.e., starting or sustaining a causal sequence that leads to foreseen harm), must include more stringent constraints, demanding more of perpetrators after the harm has been enacted, compared to reasons against allowing harm. The moral status of enabling harm—actions involving the removal of obstacles that prevent harm or the creation of obstacles to harm prevention—is another matter altogether. Doing and enabling harm share the important moral feature of giving rise to costs, while allowing harm generally does not. Giving rise to costs means that an agent's location, movements, or (in) actions consequentially lead to another agent being harmed. Giving rise to costs is morally significant; thus doing/enabling harm is morally different from allowing harm (Barry and Øverland 2016, 96–121). Agents who give rise to costs by doing and enabling harm therefore have more stringent duties to address such harm than those who merely allow harm. Big Oil's harm-related morally relevant facts thwarted harm prevention, giving rise to costs in terms of the climate harm generated. The facts are thus morally relevant and morally consistent with oil and gas companies' harm-doing—emitting greenhouse gases—and thus engender the same forms of strict duties.

Harm-doing and harm-enabling can be usefully addressed through an overarching methodology of *contribution-based responsibility*, which is part of a broader approach of corrective, or rectificatory, justice that the current analysis adopts. Corrective justice originates from harm-doing and harm-enabling and helps focus on past and present harm generated by Big Oil, elaborating on the resulting duties required to rectify such harm, consistent with the provisions of climate-related *harm-avoidance justice* (Caney 2014).

Two final specifications are in order. First, it is worth reiterating that the focus on harm generated by the companies in question in no way aims to downplay the contribution originating from the behavior of individuals, hence avoiding the accusation of *individual denialism* (Broome 2019). This perspective simply aims to shed light on the responsibility and ensuing duties of a specific group of agents, the oil majors. From a different viewpoint, the responsibility arguments developed within a corrective justice perspective do not imply that dominant burden-sharing climate ethics is wrong or impractical. Many studies have demonstrated its validity for allocating the climate burden; indisputable theoretical and experimental evidence in contexts unrelated to climate change has also long proved its effectiveness for allocating resources fairly. Rather, in relation to climate change and especially where Big Oil is the focal point of a moral analysis, burden-sharing justice seems to be incomplete. Hence, it might prove useful to integrate and strengthen this perspective of justice through an approach of corrective justice based on harm avoidance/prevention. Once this harm-based methodology to corrective justice is developed, the burden of required actions can be allocated among fossil fuel companies based on the provisions of distributive justice, as for instance chapter 10 does.

RESPONSIBILITY

As clarified in chapter 3, responsibility, especially in the context of climate change, is a composite, arduous moral concept unfortunately lacking universally agreed definitions. Responsibility can thus be interpreted, constructed, and applied in many different ways. This chapter first views responsibility within a corrective justice perspective as being in accordance with the traditional *liability model*, that is, as the condition that makes it possible to trace a situation or event back to *primary* agents who conceived it in an intentional, rational, autonomous, and morally relevant manner and to *indirect* agents who should endorse primary ones to comply with their responsibility. Then, the chapter widens the scope of responsibility to nonliability perspectives that include elements of accountability aligned with the demands posed to Big Oil by civil society.

This section, in particular, outlines the fundamental distinctions about responsibility in relation to Big Oil.

Legal Responsibility

It is imperative to distinguish between the moral and legal understandings of responsibility in the first place. Responsibility is employed in this book mostly, if not exclusively, in moral terms and not, despite its importance, in legal ones. Nonetheless, a glimpse into legal responsibility and its connection to moral responsibility is warranted. Indeed, evolving notions of legal responsibility have played an important role in social and cultural change.

For instance, in the United States, after the harmfulness of tobacco was substantiated and became common knowledge, the argument that smoking was a question of individual choice was progressively rejected, while the social acceptance of manufacturing a product that killed people rapidly decreased. This change of attitude induced the US Department of Justice to hold the tobacco industry legally responsible (and culpable) for knowingly spreading disinformation.

In the same vein, reference to the legal implications of Big Oil's responsibility is indeed useful. Ekwurzel and colleagues' (2017) and Licker and colleagues' (2019) works provide, for instance, a fundamental breakthrough in attribution science for establishing oil and gas companies' legal responsibility for climate change, as these works evince their causal contribution to climate impacts and harm. This, however, is not enough to hold agents fully responsible in a court of law. Judicial bodies also seek evidence that a defendant is culpable for the harm caused because they acted (or failed to act) in a way that renders them liable for addressing and remedying the consequences of those actions. In short, as mentioned in chapter 2, tort law and the laws on product liability rely on a general principle of ethics according to which when a clear process of causation exists, an agent must be held liable for harm if the agent had the capacity to foresee the harm and to avoid or minimize it.

In terms of legal implications, oil companies' responsibility therefore relates to their *legal liability*. Indeed, the two concepts—responsibility and liability—are inevitably intertwined: for instance, the standard legal model of liability based on contributory fault assumes that causal responsibility can be assigned to agents responsible for the dire consequences they cause through their faulty actions and consequently that they are liable for providing the appropriate remedy. Additionally, moral responsibility is the central tenet that induces remedial actions (Feinberg 1970). In this view, the fossil fuel industry's moral responsibility seems decisive for establishing its legal liability according to tort law and in terms of strict liability, as evinced by the recent explosion in climate liability lawsuits, especially in the United States. These lawsuits are expected to increasingly influence fossil fuel divestment, where moral responsibility can provide a solid enough rationale for investors to divert their business interests.

Having thus been acknowledged and clarified, responsibility for climate change raises a number of serious concerns that should be addressed through a meticulous contextual investigation in relation to Big Oil.

Conceptual Distinctions: Scope and Objectives
of the Notions of Responsibility

Responsibility can be *negative* and compel Big Oil to refrain from performing harm-generating actions, consistent with what is demanded by the do no harm principle, or can be *positive* and require that Big Oil act in certain ways. Generally, the first kind of responsibility provides the moral basis for and *triggers* the second: if an agent contributes to harm in violation of a negative responsibility, it becomes the agent's positive responsibility to redress it through immaterial, material, or financial means (Shue 2017). Additionally, responsibility can be *special* and pertain only to those directly harmed or can be *general* and be owed to all humanity and possibly to Earth. Another distinction is between *backward-looking* responsibility (which demands action based on something that has occurred in the past) and *forward-looking* responsibility (which implies that agents act because they are in a position to bring about improvements in the situation). A further crucial distinction pertinent to Big Oil is, as anticipated, the one between *causal* and *moral* responsibility. Causal responsibility can be understood as causal contribution, while a more stringent notion of moral responsibility is based on the appraisal of agents' intentions; moral responsibility assesses their knowledge, voluntariness, and control. These conceptual distinctions are important but should not be overstated, as they can tend to become rather entangled when applied to specific issues.

Big Oil's positive responsibility has roots in its negative general backward-looking responsibility; this, in turn, derives from the violation of the no-harm principle, which ought to be established in a pluralistic and nonarbitrary way to outline and justify consequent duties. This can be done by taking into account the facts of Big Oil's harm-doing, as illustrated in chapter 1 (carbon emissions associated to its processes and products) and the harm-enabling facts demonstrated in chapter 2 (awareness, behavior, capacity, denial) as well as the non–harm-based fact, enrichment. Such facts help clarify the conduct of the oil industry as well as consent an understanding of the moral context within which it operates and therefore provide a normative foundation for its composite positive responsibility and the consequent moral and practical implications.

Emissions themselves are the most blatant harm-doing fact, their mere presence indicating that Big Oil largely drove climate change by producing, distributing, and burning fossil fuels. This fact already establishes a special backward-looking, causal responsibility, which is a necessary but not sufficient condition for the more stringent notion of moral responsibility.

At the same time, Big Oil has been aware of the harmful consequences of its business model (harm-enabling morally relevant fact A, awareness) at least since the first International Panel on Climate Change report (IPCC 1990) was presented to world leaders at the Rio Conference in 1992 (indeed, there is good reason to believe that they knew some two decades before that, as clarified in chapter 2). Despite this knowledge, the majority of its harmful greenhouse gases were emitted since 1988 (harm-enabling morally relevant fact B, behavior) in a period in which oil and gas companies had the means and the know-how to limit (at least to some extent) those harmful actions (harm-enabling morally relevant fact C, capacity). Big Oil intentionally funded, shaped, and orchestrated climate denial in order to block initiatives against climate change (harm-enabling morally relevant fact D, denial). Besides these morally relevant facts evincing the backward-looking responsibility posited by the polluter pays principle, oil and gas companies have accumulated extraordinary wealth through their fossil fuel–related activities (morally relevant fact E, enrichment). As already specified, the last fact is not related to harm per se; it is nonetheless an account of backward- (prompted by the beneficiary pays principle) and forward-looking (prompted by the ability to pay principle) responsibility that strengthens the cogency of Big Oil's overall moral responsibility.

Based on these morally relevant facts and consistent with the possibility of a notion of moral responsibility for collective entities, Big Oil must be held morally responsible for climate change. Specifically, these facts justify assigning positive, special, backward- and forward-looking, and moral responsibility for climate change to oil and gas companies. Such responsibility is a normative construct focused on oil and gas companies' conduct and intentions in the context of the violation of the no-harm principle, which also takes into account their extraordinary wealth; their composite responsibility provides the moral basis for the establishment of duties that compel them to act in certain ways. These duties should be understood as *informal sanctions* imposed by the moral nature of oil and gas companies' composite responsibility (Jamieson 2015).

MORE POLITICAL AND PRACTICABLE NOTIONS OF JUSTICE AND RESPONSIBILITY FOR BIG OIL

To move toward more political and practicable notions of justice and responsibility for Big Oil, it is first necessary to take stock. This book has argued that corrective justice can provide the opportune normative construct to uphold claims of responsibility for climate change against Big Oil as well as for framing the consequent duties. Corrective justice entails a positive, special, backward- and forward-looking moral responsibility, which largely responds to the logic of *imputability*, ascribing fault-based liability-responsibility for climate harm in the form of duties.

Such theoretical notions of corrective justice and responsibility should be integrated with broader and more political and pragmatic conceptions, rooted in the demands for climate justice made by civil society stated at the outset of chapter 3: to rectify the harm already done and to discontinue fossil fuels in order to avoid further future harm. Such demands address the structural injustice of the climate crisis arising from the carbon-intensive structures, practices, and institutions that constitute the global political and economic system (Sardo 2020). Consequently, they should prompt an inclusive effort to arrive at possible solutions to right previous wrongs and to ensure that humanity's energy needs in the future are innocuous to the planet.

The objective of putting forward broader and more political notions of justice and responsibility is twofold: on the one hand, it responds to the

necessity to strengthen the theoretical dimensions of the quest for justice in climate change; on the other hand—and more importantly in this
context—wider-reaching and more politically relevant notions of justice
and responsibility may provide the connective tissue between the theoretical and the practical dimensions of climate ethics, thereby facilitating
and/or increasing the political and institutional real-world influence of its
moral tenets in relation to Big Oil. In particular, to move toward greater
practicability and effectiveness of the theoretical constructs outlined and
closer to the demands posed by civil society, the scope of justice must
be extended to recognition and participation and that of responsibility
to accountability. This extension of the notions of justice and responsibility helps transform the usual top-down approach of *government* to a
more bottom-up approach of *governance*, consistent with the aforementioned requests of climate justice hailing from civil society and capable of
accommodating the potential proactive role of indirect agents of justice,
thus acting as a countermeasure to the increasing lack of legitimacy of
traditional government-based policy making.

Justice as Recognition and Participation

Distributive and, to a lesser extent, corrective justice provide the life-
blood to academic literature on environmental and climate ethics. In this
perspective the environment is a resource that should be governed by
principles of justice. Distributive justice deals with the distribution of the
benefits and costs of environmental resources; corrective justice is about
punishment and compensation for the harm caused to others by wrongfully appropriating or using environmental resources.

In fact, climate justice movements—which largely emerged from the
anti–carbon markets movements of the 1990s—broadened the focus
to the political economy of climate governance, demanding something
more than mere distributive and corrective justice. For instance, these
movements aimed to shift socioeconomic systems to a postcarbon world
and make those responsible for the ecological and social damage of climate change pay, demanding that the voices and interests of the most
vulnerable peoples and communities be taken into account. On the

whole, they largely focus on changing the nature of a harmful yet powerful production system, requiring financial rectification for harm done and providing the possibility of participating in decisional processes autonomously. In other words, global environmental/climate justice movements aim at broadening the scope of justice to political issues embracing all peoples and communities affected and to participation in the political processes that create and manage the different political approaches to the environment.

There are multiple dimensions and nuances to any discussion of environmental and climate justice coherent with *social justice* aims; one of these is undoubtedly a consideration of the impacts on future generations. These multifaceted perspectives on justice call into question the dominance of the *distributive paradigm*, highlighting issues of recognition, difference, and political participation in a model where justice and responsibility are socially connected (Young 2006).

The concept of recognition—intended both as basic respect of and meaningful engagement with diverse values, cultures, perspectives, and worldviews—is crucial for eliciting and coalescing the different claims of climate justice emerging from civil society and environmental movements that point the finger at Big Oil. Indeed, recognition pays particular attention to populations and groups that have been excluded or marginalized in climate policy and governance and to those who are particularly vulnerable to climate change. This is especially pertinent in the case, for instance, of oil and gas companies in less developed countries, where they often have a history of backing and funding authoritarian regimes with the objective of exploiting fossil fuels through processes that have never taken into consideration the needs, vulnerabilities, and interests of local peoples and communities, the legitimate owners of such resources (Wenar 2011).

Participation is another fundamental component with regard to environmental and climate issues. Taking part in decisional processes involving one's own interests guarantees the fair distribution of rights. In the case of oil and gas companies, these processes include the entire range of activities tied to fossil fuel production. Indeed, a lack of recognition necessarily generates poor participation; this is largely witnessed in terms of how race, class, and gender remain structural barriers that marginalize individuals and entire communities. In this regard, environmental and climate movements call for decisional processes that would break down the structural and cultural obstacles, embracing cross-cultural formats and exchanges to enable the participation of diversities. Such decisional processes were completely overlooked by Big Oil; the industry made conscious choices in settings impenetrable to stakeholders, choices greatly deleterious to the health and interests of the wider public.

Political and Pragmatic Responsibility

As anticipated, the standard model of responsibility grounded in corrective justice can be assimilated with a *liability model*, which is mostly backward-looking and based on the causal chain connecting agents and events (Pellizzoni and Ylönen 2008). This form of liability-responsibility imposes duties on oil and gas companies, compelling them to undertake (costly) rectificatory actions. Even if they can delegate some of these costs, spontaneous compliance to their duties is not a given. Consistent with the broader and more inclusive notion of justice as recognition and participation, it seems instead, as said in the ensuing section, more viable to imagine that compliance will be achieved through (second-order duties imposed by) second-order responsibility, that is, the responsibility on the part of indirect agents who should undertake a number of tasks to ensure that first-order agents comply with their (first-order) duties.

To influence the ebbs and flows the oil industry faces, made up of opportunities, constraints, and incentives, the moral relation between Big Oil—the first-order agent—and indirect agents must be based on a broader notion of responsibility, one that includes more empirically grounded considerations and is able to impact politics and to be more workable in the long term. This ensures that it can usefully complement the liability model of corrective justice while being attentive to elements of recognition and participation.

Whereas a liability-based account of responsibility demands a strong authority able to improve and govern states of affairs, a political and pragmatic notion of responsibility is closely linked to accountability and demands a different relationship, one that goes beyond the limits of traditional imputability and includes the demands of various stakeholders. Accountability is a core principle of good governance that transforms responsibility from simple imputation to reasoned justification (Pellizzoni 2004). In short, accountability is the willingness to account for one's actions and has two key dimensions: answerability and enforceability (Grant and Keohane 2005).

Answerability means that some agents have the right to hold other agents to specific standards and to appraise whether they have fulfilled the achievement of these standards; enforceability implies that the former can impose sanctions if they determine that these standards have not been met. In this purview, Big Oil is answerable to indirect agents who have the right to hold oil companies to the set of standards required by the duties of reparation and decarbonization in order to fulfill their responsibility to do no harm. Additionally, given that they continue to do harm through their fossil fuel–related activities, the sanctions that indirect agents impose on Big Oil must aim at minimizing such harm.

Such demand of answerability and, more broadly, the pragmatic understanding of responsibility required by the notion of accountability pave the way for the analysis of second-order responsibility and duties in relation to Big Oil.

SECOND-ORDER DUTIES VIS-À-VIS BIG OIL

In light of the nuances of Big Oil's responsibility and of its extension to a more political and practical notion, a distinction that—given its importance in this context of analysis—needs to be addressed thoroughly is between *first-order* duties, or the duty to perform or omit certain acts (to which the book has referred to in general terms thus far), and *second-order* duties, or the duty by indirect agents to ensure that first-order or direct agents comply with their (first-order) duties (O'Neill 2001, 2005; Caney 2014).

It should be noted that an allocation of second-order duties to indirect agents risks being ineffective: what if, similarly to first-order agents, indirect agents too fail to act on their second-order responsibility? This shortcoming, however, largely depends on reasons stemming from the indeterminacy of the allocation of second-order duties. As emphasized at the outset of chapter 3, to obviate this indeterminacy risk, such allocation should be supplemented with specific and localized information sourced by engaging with the actual politics of climate action in order for it to be applied by indirect agents to actual situations. This context dependency thus increases indirect agents' motivation and willingness to pursue change, as it realistically indicates a way for them to have a positive influence in the real world (Green and Brandstedt 2021). The moral framework developed details within both the duties of reparation and decarbonization the specific courses of action that particular indirect agents should undertake if they want to compel Big Oil to address the harm associated with its business or at least steer the industry in the right direction, as evinced in general terms in the remainder of this section and discussed in chapter 8 in reference to the actual politics of Big Oil in a world shaken by the climate crisis.

Responsibility imposes duties that must be realized through practical actions; most of the time, such rectificatory actions require some sacrifice. Oil and gas companies must modify their behavior and should be required to financially rectify the harm done and decarbonize their business, as explained in the rest of the book. This entails major outlays of cash but, broadly speaking, must also include nonmonetary burdens, such as opportunity costs. Even if the industry can pass some costs on to other agents (e.g., consumers, governments, other businesses), it is questionable whether they would spontaneously comply with their duties (this is a shortcut used in what follows, meaning *to comply with the behaviors and actions imposed by duties originating from responsibility*). As already emphasized, the motiving reason to take action would be in all likelihood insufficient. Therefore, realistically, it is necessary to employ the distinction between *first-order duty* (the duty to act) and *second-order duty* (the duty to ensure that other agents act).

A first option would be that, in the event of lack of compliance by the designated agent, other agents would cover the noncomplier's duties (Shue 1996, 71–73). This option, albeit plausible, is to some extent already occurring. The current state-centric international climate regime under the United Nations Framework Convention on Climate Change largely discharges subnational agents' duties onto states and, in the present analysis, appears insufficient because it would only be a reactive response to the nonfulfillment of duties by Big Oil. Additionally, this option would be morally unacceptable since it would irremediably undermine the entire case for the unique agency of Big Oil in climate change; therefore it will not be taken under examination here.

A more nuanced solution for strengthening agents' compliance is the one provided by Caney (2016a, 2016b). He argues that when some agents do not honor their (first-order) duties (and responsibility), the shortfall can be addressed in five ways:

1. by aiming for less ambitious targets,
2. by allocating a portion of said duties to other agents (as suggested above),
3. by sharing part of the burden with other agents,
4. by including other moral ideals beside justice, and
5. by changing the incentive structure of the contexts in which agents operate.

Yet, options 1, 2, and 4 are still reactive, whereas option 3 is only mildly proactive, since at most it aims at changing agents' behavior by inducing them to act in ways they would otherwise have avoided. Altogether, given Big Oil's role and distinctive agency in the current climate order, the approaches proposed in options 1 to 4 do not seem sufficiently adept at increasing the likelihood of compliance with its first-order duties in a significant manner. As anticipated, the approach to compliance outlined in point 5 is more appealing for the argument developed in this book, whereby Big Oil (the first-order agent) is induced to comply with first-order duties through the actions influencing the opportunities, constraints, and incentives of the social, economic, political, and legal contexts it operates in demanded to other agents by their second-order duties.

The most important elements of this account of second-order duties are the *tasks* (i.e., what needs to be done to minimize climate harm in this case) and the *actors* (the indirect, or second-order, agents) most suited for carrying out the tasks (Caney 2005, 2014, 2016a). In fact, given their practical relevance, such indirect agents are subsequently framed—in part III of the book, which has a more empirical approach—as *agents of destabilization* of the status quo in which Big Oil thrives.

By *matching* tasks with actors, it is possible to develop a full account of second-order duties. Six tasks for averting climate change and its related harm could be contemplated in relation to Big Oil, schematized in figure 4.1:

1. *Legal framework*. Establishing the legal and political framework of climate governance for supporting the realization of Big Oil's duties.
2. *Enforcement*. Agents with the political power of creating enforcement mechanisms, including transparency ones, have an obligation to do so.
3. *(Dis)Incentivization*. The cost of fossil fuels can be increased by cutting part or all of the staggering subsidies the oil industry receives, by making consumers pay the full cost of carbon, or both. Alternatively, producers and consumers can be incentivized to switch to cleaner sources of energy.
4. *Enablement*. Facilitating scientific research (into clean technologies, new energy sources, energy efficiency) and transferring scientific innovations widely, also in the developing world, to enable the low-carbon transition.
5. *Spreading of norms and practices*. Fostering and maintaining social/moral norms and good practices that, for instance, support recognizing

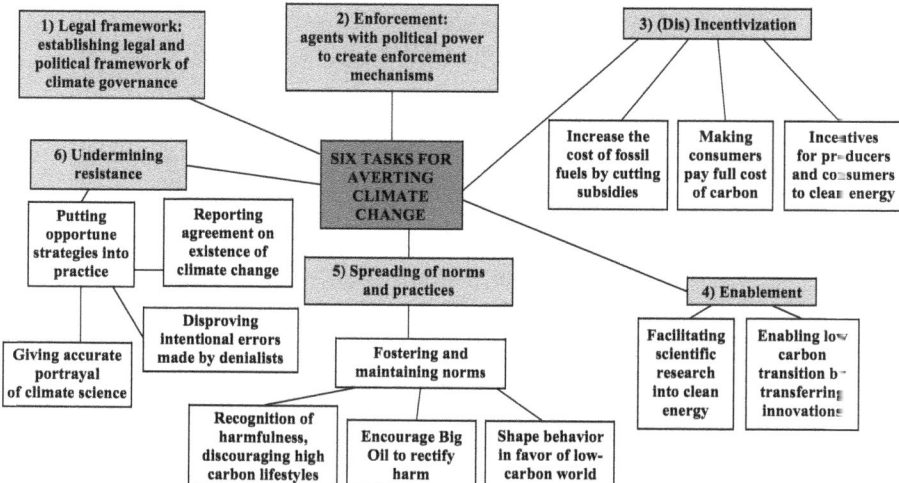

Figure 4.1
Tasks for averting climate change.
Source: Author.

the harmfulness of fossil fuels, discourage high-carbon lifestyles, encourage Big Oil to rectify the harm done, and, more generally, try to shape behavior in favor of a less harmful low-carbon world.

6. *Undermining resistance.* Most of the time it is denial and power that have produced resistance to climate action (see chapter 2 and part III). To undermine such resistance, opportune strategies must be put in practice, such as giving an accurate portrayal of climate science, reporting the levels of agreement on the existence of anthropogenic climate change, and disproving intentional errors and misinterpretations made by denialists.

Who are the actors most suited for carrying out such tasks or, according to the vocabulary adopted, the indirect agents responsible for promoting Big Oil's compliance to its first-order duties? As previously suggested, this depends greatly on the task at hand: some indirect agents such as political authorities (governments) and international organizations, as would be expected, play a crucial role. However, other less obvious agents can also give noteworthy contributions in the current context of analysis.

For tasks 1 and 2—legal framework and enforcement—it might seem that only international, national, or subnational political authorities can establish the legal and political framework of climate governance and policy for Big Oil to enact its (first-order) duties. In many respects, however, citizens and, more generally, civil society can play a role by ceasing to support political authorities that do not endorse actions for adequately contrasting climate change. Given that virtually all major—privately or publicly owned—oil and gas companies' activities have international reach, national frameworks need be to be collaborative affairs, a process that could be underwritten by international organizations such as the United Nations Framework Convention on Climate Change, the Global Environmental Facility, the World Trade Organization, and the World Bank, to name but a few players with worldwide stature.

Task 3—(dis)incentivization—remains under the control of governments and other subnational political authorities, but in this instance too, international organizations play a fundamental role, as they can endorse strategies for steering the actions of Big Oil through international frameworks of collaboration.

The remaining tasks suggest less conventional indirect agents. Enablement—task 4—is largely accomplished through research, innovation, and the diffusion of knowledge and information. In this case, research councils, universities, research centers, and innovation agencies all play a key role. Political authorities and international organizations may rubber-stamp these agents through channeled funding, promotion of collaborative agreements, and so on.

In the case of task 5—social/moral norms and practices—the relevant agents veer off the path of the more conventional large-scale figures of authority. A significant role can be played by figures such as religious leaders, intellectuals, communicators and influencers, and other charismatic individuals. They should support and diffuse social norms that hold fossil fuel–related activities as being morally wrong.

Similarly, for task 6—undermining resistance—the most suitable agents are, on the one hand, those who can most faithfully, reliably, and effectively communicate climate science to the public: climate scientists able to speak in layman's terms, science journalists with mass media appeal, other reliable investigative media sources, and environmental advocacy groups such as the Union of Concerned Scientists and the Center for International Environmental Law. On the other hand, political authorities at various levels (national and subnational entities) and economic agents (banks, insurers, and asset management funds) all have a vital contribution to make.

This account of second-order duties has important normative and practical implications for Big Oil. On normative grounds, the discussion sheds light on actions that differ from those usually cited for coaxing Big Oil into fulfilling its (first-order) duties of reparation and decarbonization and, furthermore, looks beyond governments to identify a wider range of indirect agents, thus using a different normative basis for ascribing second-order duties. In practical terms, by matching tasks with actors, it becomes clear that several diverse indirect agents can in different ways contribute to modify Big Oil's behavior, thus favoring compliance with its duties. These indirect agents and their role in destabilizing the status quo in the industry will come under closer scrutiny in part III of the book.

In the immediate aftermath of the International Panel on Climate Change's (IPCC) publication of the landmark *IPCC Special Report on Global Warming of 1.5°C* (IPCC 2018) urging the rapid phasing out of fossil fuels to avert the direst consequences of global warming, Shell CEO Ben van Beurden affirmed to industry leaders at the Oil and Money Conference in London in 2018 that a huge tree-planting project the size of the Amazon rain forest would be needed to achieve the 1.5°C target. Is there anything intrinsically deplorable about this claim besides its patent impossibility, as the finite space on our planet would not allow for such mass-scale tree-planting projects, roughly equivalent to a small continent (ActionAid et al. 2020)?

According to the analysis in the previous chapters of part II exploring Big Oil's role in the climate crisis from a moral perspective, van Beurden's words are an attempt to redefine the parameters of the morally acceptable, as they are a testament to Shell's willful disdain toward its responsibility arising from the harm done by the fossil fuels manufactured by the company. Such obliviousness makes Shell implicitly reject across-the-board duties of addressing climate harm that moral reasoning assigns to it, despite its repeated green claims and pledges. Shell's plan appears, rather, to assume the form of an attempted *easy way out* through actions (tree-planting) involving unspecified agents that would have to relinquish something (land, labor, investments, time) to avert the dire situation that the IPCC report depicts. And talk of trees always garners approval from the marketing spin doctors, the media, and the wider public. Unfortunately, Shell and their fossil-fuel–producing peers cannot shelter from their moral responsibility under the protective canopy of a forest.

Indeed, the Shell CEO's claim prompts a host of questions on the controversial connection between the oil industry and society, which has been a love-hate relationship since its very inception. In the good old days it was comforting to believe, among the many reassuring tales of that not-too-distant golden age of hope and abundance, in the industry's "*Happy Oil*" narrative. It basically trumpeted that what was good for the oil industry was good for all, since any human problem could be solved through more fossil fuels and fossil-based technological breakthroughs. The immediate advantages of a plentiful source of cheap oil were extraordinarily compelling. *The Guardian* newspaper painted an insightful portrait: "the fossil fuel industry told us that we could take out an interest-only mortgage against the future of the planet and prices would always go up, interest rates would always go down and there would never be a reckoning" (McDuff 2018). But like the 2008 subprime mortgage crisis that led to one of the most severe financial crashes since the Great Depression, things had to change, and the image of the *Happy Oil* bubble burst dramatically. Humanity now knows that the Faustian pact made with what went on to become known as Big Oil was not actually that advantageous, and, in fact, "we [humans] find ourselves facing repayments on the scale of trillions of dollars. That does not even cover the human costs that these dry figures obscure: the lives lost, the homes flooded, the farms wasted away to drought."

This chapter explores some salient features of the complex relationship between the industry and the society it operates in to further clarify how Big Oil's responsibility for climate change can be translated into duties. It is worth recalling that these duties of reparation and decarbonization are, so to speak, the bridging elements that connect the more normative-theoretical perspective of responsibility examined in chapter 4 to the more political-empirical perspective of climate governance and politics aligned with the demands of civil society covered in part III.

To magnify this interlocking role of duties, it is worthwhile to further investigate a few elements of Big Oil's conduct in relation to society. In particular, this chapter first defends the claims of duties owed to society against the most pressing counterclaims that dispute this view: rebuttals in which the terms *benefit provision, consumption-based,* and *law-abiding* ring loud. Then, to fully clarify the relationship between Big Oil and society, the chapter focuses on the moral status of fossil fuels per se and on the intrinsic moral wrongness of the harm the industry has caused. The chapter goes on to point out some specifications of Big Oil's duties in light of the demand emerging from society. Finally, the chapter sheds some light on how Big Oil has started to respond to society's expectations, especially in relation to decarbonization.

COUNTERING THREE COMMONSENSE REBUTTALS

In order to foster the cogency of Big Oil's duties toward society, it is necessary to systematically argue their case against the most pressing claims that dispute this view. Basically, this section addresses three common and apparently sound rebuttals to oil and gas companies' responsibility and duties, often used by fossil fuel advocates: (1) thanks to their fossil fuel–related activities, oil and gas companies have largely benefited humanity; (2) final consumers of fossil fuels are the ultimate agents responsible for emissions; and (3) oil and gas companies, at least in democratic societies with well-functioning markets, have no other obligations beyond what is required by law.

The Benefit Provision Rebuttal

Fossil fuels have benefited societies and improved the quality of life of humanity: given this, some observers point out that all things considered, associated costs must be tolerated. The more ardent supporters of this thesis loudly proclaim that the benefits far outweigh the costs.

There are a number of reasons why the cost/benefit frame should be avoided in analyzing the role of Big Oil in climate change. First and foremost to consider is that the moral obligation to *do no harm* takes precedence over any cost/benefit considerations, and the actual harm being done overrides the concept of any societal gains. Second, any benefits that were obtained by society at large were not deliberately premeditated by Big Oil, whose only intention was to reap as much of a financial reward as possible. Intentionality is a fundamental issue to defining *moral* responsibility: that the fossil fuel industry provoked positive—although unintentional—repercussions on global wealth and society is undeniable but irrelevant to the question of moral responsibility. Indeed, Big Oil has been more than amply rewarded for any benefits to society by the extraordinary wealth accrued over the years, so this aspect should not counterbalance the costs in any cost/benefit consideration.

Parallels can be drawn with the past. Mainly thanks to the world trade in cotton, the antebellum South in the United States was one of the economic engines of a thriving country. Its fuel of choice? Slavery. The same was to some extent true of the British Empire; both economies could claim that slavery had enabled them to produce great benefits for their industries and therefore their citizens. And while lawmakers and religious institutions of the day were split on the moral question of human bondage and its inhumanity, there remained consensus over its economic necessity. Today, nobody would argue that a cost/benefit analysis could ever justify such atrocity. Social norms eventually adapted, favoring the growth of a large abolitionist movement and finally leading to the establishment of antislavery laws before being formally abolished by the US Congress in 1865.

And yet some of the very same rhetoric used to defend slavery is adopted today to endorse the use of fossil fuels. In his *Memoir on Slavery,*

South Carolina senator William Harper (1838) made an appeal against a rash overturn of the slave-centered economic status quo that would upset the pecuniary interests of the South: "Very different indeed is the course of [the abolitionists] whose precipitate and ignorant zeal would overturn the fundamental institutions of society, uproar its peace and endanger its security, in pursuit of a distant and shadowy good, of which they themselves have formed no definite conception—whose atrocious philosophy would sacrifice a generation—and more than one generation—for any hypothesis" (qtd. in Davidson 2008, 73). Replace the word *abolitionists* with *environmentalists*, and they are barely distinguishable from the sentiments expressed by today's oil lobbyists.

But the cost/benefit debate is also misleading on a methodological, so to speak, front: it only takes a short- and medium-term view. A long-term cost/benefit approach would inevitably have to take into account the infinitely higher costs of a climate catastrophe and the related economic collapse.

The Consumption-Based Rebuttal

Final consumers, through the choices they make on the marketplace, are indeed the main agents who shape and determine the demand for oil and gas. So, why hold the industry responsible and duty-bound for the emissions generated by the consumption of oil and gas if they simply meet a demand for those products freely expressed by autonomous agents on markets or otherwise?

At the very least, it is dubious to point the finger at consumers as the sole or even main agents responsible for climate change in that, by and large, they have a limited ability to express their proenvironmental preferences. Consumers have less information than producers on the negative externalities that fossil fuels provided by the latter generate: in the case of carbon emissions, consumers are, for example, typically unaware of the long life span of greenhouse gases in the atmosphere, whereas Big Oil's scientists and executives have been aware of the facts for a long time. Furthermore, current consumers are also somehow *culturally trapped* in fossil fuels: most of us grew up in a time when fossil fuels were *the good thing* and when there was no problem that could not be solved by throwing more fossil fuels at it. It is certainly a challenge to modify these entrenched mindsets.

From a different perspective, *greenwashing* has become pervasive in the oil industry. This practice involves putting an environmentalist spin on marketing to deceive the public into believing in the industry's green credentials; many consumers therefore do believe that fossil fuels are becoming less disruptive for the planet and humanity. Additionally—and this is not a minor issue—many consumers lack the access to or the financial capacity to afford less carbon-intensive goods and services.

Even in the absence of such distortionary conditions, markets do not reflect the environmental and social values of individuals, who are in fact not only consumers. As consumers, individual agents are forced to make choices that differ from those they would make as citizens (Sunstein 1997): limited alternatives, material impossibility, routines, and habits can lead them to knowingly utilize carbon-intensive products despite an inclination that would otherwise lead them to make alternative choices. Therefore, as emissions are dictated by dynamics largely beyond individual control, the same individuals are likely to experience a sense of impotence (Cuomo 2011). As a consequence, it is even more psychologically problematic for disenfranchised individuals to reduce their externally constrained emissions.

This consumption-focused standpoint, besides obfuscating the responsibility and role of Big Oil, establishes a sort of *personal sacrifice trap* that ignores that consumption choices are constrained by a complex sociopolitical and regulatory landscape and powerful economic interests that promote the use of fossil fuels. Individual actions are surpassed by the structural dynamics (Bernstein and Hoffmann 2019) of the oil regime, as evinced in part III. Focusing only on demand for oil and gas would attribute the failure to address climate change solely to consumers' lack of green credentials while obscuring the sociopolitical and regulatory structures that shape their choices (Lenferna 2018). Furthermore, such an emphasis is counterproductive, as psychological investigations show that a focus on consumers' personal responsibility decreases the individual's willingness to engage in proenvironmentalist behavior (Obradovich and Guenther 2016; Lavallee et al. 2019).

Such individualistic rhetoric—pointing the finger resolutely at the consumer as sinner—is, in fact, another consequence of decades of oil industry propaganda and has contributed to undermining climate action (Supran and Oreskes 2021). Big Oil's deceptive narratives have succeeded in framing the question of climate change as one of individual consumption-based responsibility, thus preventing the general public from understanding that the climate crisis is a *structural* problem largely driven by the oil industry's denial, misinformation, lobbying, and disablement of climate policy and legislation. For instance, the notion of the personal carbon footprint was first popularized by BP: in its "*Beyond Petroleum*" rebranding, the oil major in the early aughts introduced and promoted the term *carbon footprint* and launched one of the first personal carbon footprint calculators that provided untold hours of fun at dinner-table conversations.

In this way, Big Oil has managed to circumvent responsibility and duties for climate change and to present itself as a mere supplier of a product that meets existing demand rather than as the major underlying cause of the problem: these *deflection campaigns* have long been funded by industry (Mann 2019). Contrary to widespread belief, Big Oil has always been keen to discuss climate change as long as the dialogue was kept at the level of individual responsibility and duties—on light bulbs, single-use straws, or exotic fruits shipped from around the globe—nimbly avoiding reference to its own responsibility or systemic change.

It is worth recalling that final consumer emissions, especially those of individuals, make up a minor share of total global emissions: this is known as *the insufficiency problem* (Cuomo 2011). This problem does not imply that consumers' efforts to cut their emissions are not justified; indeed, especially in regard to wasteful and unnecessary/luxury emissions, consumers do have a duty to limit such emissions and shift their consumption patterns toward less carbon-intensive goods and services. Dismissing individual emissions altogether as negligible would be a mistake in moral mathematics. Individuals at this point in time must know that the emissions generated by their consumptions are part of a bigger picture of similar actions that together result in greater harm, so this should be their cue to make informed choices about their consumption-related emissions.

This is not, of course, an attempt to alleviate individuals from any personal responsibility and duties or discourage their meaningful personal engagement in the fight against climate change, given that even scientific evidence shows that individual lifestyles—particularly those of climate change communicators—do have a remarkable systemic impact (Attari, Krantz, and Weber 2019). At any rate, consumer responsibility and duties to curb/stop their emissions are small in comparison to those of large corporate emitters. Therefore, while individual consumers have a duty to take meaningful and adequate action to reasonably limit their emissions—especially the better-off ones whose lifestyles obviously cause greater emissions to the point that the richest 1 percent produces more than twice as much emissions of the poorest half of humanity (Oxfam 2020)—they should also feel duty-bound as indirect agents to be a driving force in imposing duties on *metalevel emitters,* among which oil and gas companies are prominent first-order agents.

The Law-Abiding Rebuttal

While it is unanimously acknowledged that Big Oil must take part in the global struggle against the climate crisis if this effort is to be successful, it seems less obvious whether oil and gas companies have further obligations besides those set by legislation.

To investigate this issue, the shareholder versus stakeholder debate of business ethics (also known as the *Friedman-Freeman debate* from the names of the two main scholars defending the two different positions) must be briefly addressed. In a nutshell, the first view holds that corporations must focus on the interest of shareholders, while the second holds that corporations must consider and balance the interests of a wider group of stakeholders (Arnold and Bustos 2005; Hormio 2017). The 1976 Nobel laureate in economics Milton Friedman, father of the managerial shareholder theory, affirmed that business, as long as it plays by the rules imposed by the society it operates in, need only concentrate its activities on whatever will increase its profits (Friedman 1962). He maintains that managers have a legal fiduciary duty that requires them to make decisions on behalf of the corporation to further the interests of shareholder and that the social responsibility of business is to increase its profits (Friedman 1970). On the other hand, the stakeholder theory, while not denying that profit is a necessary goal of business, argues that its primary objective is to manage stakeholder interests (Freeman 1984). Freeman christens the view that business decisions should be considered as distinct from ethical ones as the *separation fallacy*, whereby profiteering is free of any moral dimension, as it is merely a business decision. The stakeholder theory, aligned with the demand of corporate social responsibility, requires executives to pursue goals that go beyond the mere interests of shareholders, practicing the interests of a broader constituency of societal stakeholders.

The overall merits of these two views will not be assessed here. For our purposes, suffice it to say that in the shareholder theory, endorsed by and large by economists and business schools but increasingly called into question outside this perimeter, abiding by laws generated by the democratic process grants full legitimacy to corporations' actions. Market failures, such as greenhouse gas emissions, are addressed through regulations and economic instruments; the absence of any such regulations presumes that citizens voted against imposing any such directives, so corporations have no responsibility whatsoever for their lawful processes and products emitting greenhouse gases. Thus, the shareholder theory would suggest that the harm generated by them is morally permissible (Bowie 2013).

It seems, however, simplistic and somewhat instrumental to exclude corporations from climate change responsibilities by claiming that regulations and market-based solutions in democratic contexts are open and revisable processes. First and foremost, as part III clearly shows, although most oil and gas companies—at least international oil companies (IOCs)—are based in democratic contexts, they are nonetheless part of a hegemonic bloc with political authorities, governmental agencies, influential segments of civil society, media, and the epistemic communities with the capacity to shape and establish policies, norms, and institutions that structure the climate governance system in ways that are sympathetic to their interests (Levy and Newell 2002). From a different perspective, given the overwhelming scientific consensus on the relation between carbon emissions and climate harm, the precautionary principle alone would justify corporations taking resolute action to mitigate climate change.

In particular, there are further unique elements that make Big Oil's inclusion among, so to speak, *proactive* main agents of climate justice inescapable. First, given its much greater expertise than other constituent groups with regard to the dynamics of climate change, Big Oil—all else being equal—should have greater responsibility and duties in addressing climate harm (Bowie 2013) through, for instance, developing and promoting less carbon-intensive alternatives, taking steps toward becoming Big Green Energy. Second, as underlined in chapter 2, at least some of the major IOCs have contributed to climate change by funding, shaping, and advancing climate denial. By engaging in these activities and campaigns, the corporations have stepped outside their normal sphere of influence, wielding their power in places that in normal circumstances should be alien terrain, such as in national and international policies and treaties trying to combat climate change (Hormio 2017).

Furthermore, besides denying its existence and/or its severity, the role played by anthropogenic carbon emissions, and its science, one of the main objectives of Big Oil's underhand struggle with climate change was impeding and/or slowing action to address it. As said, leading investor-owned oil companies actively opposed and in many cases successfully prevented policies to reduce greenhouse gas emissions. In brief, through such behavior, these companies have undermined the authoritativeness of the entire oil world and paralyzed global climate policy for decades. Big Oil's denial and political disablement activities cannot therefore be justified through the dominant business ethics normative view of market and business practices in democratic societies.

To sum up, this book espouses a view consistent with the stakeholder theory that industry has a greater obligation to protect the environment than the obligations established by laws and that it must also develop and demonstrate environmental moral leadership. Given the facts previously examined, this is particularly true for Big Oil, which should respond to the needs and pressures of society and coalesce into an increasing involvement in environmental and community concerns if it wants to retain its *social license to operate*, that is, the ability to meet the expectations of society and avoid activities that societies consider unacceptable (Gunningham, Kagan, and Thornton 2004).

The invalidation of the law-abiding rebuttal further confirms that Big Oil is morally responsible for climate change, and, as a full-fledged agent of climate justice, it has overwhelming duties to make amends for the harm done and to decarbonize its business in order to avert any further harm.

THE MORAL WRONGNESS OF FOSSIL FUELS AND THE EMERGENCE OF ANTI–FOSSIL FUEL NORMS

To fully fathom the complex relationship between Big Oil and society, it is necessary to investigate the moral status of fossil fuels per se in view of dealing with the intrinsic moral wrongness of the harm caused by the former to the latter. This is a further harm-centered perspective that adds to the moral salience of the harm generated by fossil fuel combustion. It is, in fact, the overall moral wrongness of dealing in fossil fuels that in many respects prompts the more practical demands of justice and responsibility that Big Oil will have to address in terms of the duties of reparation and decarbonization.

Humanity has sourced far more fossil fuels than it can safely burn. Welsby et al. (2021) claim that by 2050 nearly 60 percent of oil and gas and 90 percent of coal reserves should remain unused in order to meet the 1.5°C target. The upshot is that despite the enormous importance of fossil fuels for almost every aspect of human life, most must remain unburned and be substituted by low-carbon/noncarbon–emitting resources.

In an interview with *The Telegraph* in 2000, the former oil minister of Saudi Arabia, Sheik Yamani, famously proclaimed that "the Stone Age came to an end not because we had a lack of stones, and the oil age will come to an end not because we have a lack of oil" (Fagan 2000). Basically, Yamani claimed that cost issues and technological improvements would sound the death knell for fossil fuels, with cheaper and more effective clean energy a metaphor for the Bronze Age tools that provided the nail in the coffin of the Neolithic period. Yamani's was a rational claim, yet it seems incomplete: in order to bring about changes in the political and social conditions to end the *Oil Age*, economic and technological rationales must be complemented by social/moral norms that hold fossil fuel–related activities morally wrong because of the harm they cause. Such norms would in fact provide the theoretical background for initiatives aimed at destabilizing the oil world based on the highlighted claims of civil society about the impacts and emissions debts, such as those envisaged in part III.

A norm is here understood as a given behavior expected of a particular agent and usually enforced through social sanctions. An anti–fossil fuel norm should aim at changing the behavior of the entire industry in line with relevant conceptions of responsibility and duties. For such norms to work, people must be convinced, intellectually and emotionally, that the relentless use of fossil fuels poses a danger to humanity and Earth, and therefore it is morally wrong to continue their use. Such norm-based models would consist of the direct imposition/prohibition of select actions (rectifying the harm produced by climate impacts and operating with fossil fuels) and in forms of suasion aimed at inducing agents to modify their behavior. More operational details of anti–fossil fuel norms are provided in part III.

What are the moral reasons that can prompt the emergence of anti–fossil fuel norms? To address this question it is useful to first clarify the notion of moral progress: it is widely believed to consist of the dominance of objective/impersonal reasons for action over subjective/personal ones (Buchanan and Powell 2015; Jamieson 2017). In this view, moral progress is pragmatic and strictly context-dependent. Accordingly, limiting climate change is exemplary moral progress for the current times. This objective requires maximizing current efforts, a herculean achievement that can be carried out only by restraining and eventually abandoning fossil fuels. In short, anti–fossil fuel norms can emerge only if fossil fuels are widely acknowledged as being products that are morally wrong and with Big Oil having both responsibility and duties. Accordingly, the phasing out of fossil fuels would be moral progress for humanity.

At the root of the moral wrong caused by fossil fuels is, as said, the harm that their combustion brings about, yet the very act that causes damaging carbon emissions has always been and still is seen as natural, necessary, and inevitable in our lives (Jamieson 2017). Some scholars compare the naturalized use of fossil fuels to the passive acceptance and, in some quarters, endorsement of slavery before its abolition (e.g., Davidson 2008; Mouhot 2011; Hayes 2014; Jamieson 2017): both have major roles in sustaining economic growth while causing untold harm. The same reactionary rhetorical arguments—the ruinous economic effects their banning would cause—that defended the appalling use of slave labor are now recycled to defend the continued use of fossil fuels. Indeed, it is mostly in these arguments that the naturalization of the two is embedded, with each branded as a necessary cornerstone of the status quo (Jamieson 2017).

While this book will not investigate the (certainly instructive) comparisons between slavery and fossil fuels addressed by the authors cited above, the main upshot of their analyses is crucial: in order to increase the emergence of a social/moral norm against fossil fuels, it is first and foremost necessary to *denaturalize* their use. Step one in this challenge is to loudly proclaim the harm they cause to humanity and the environment, reiterating it time and time again. This is an arduous undertaking, though, despite the extraordinary progress in attribution science. As said in chapter 4, harm arising from climate change is a distant and abstract one—an impersonal harm—that makes it easy to argue that humans tend not to see climate change as a moral problem and therefore they lack the motivation to act with the urgency of usual responses to moral challenges.

So, despite the fact that recent studies—for instance those already cited in previous chapters by Heede (2014), Ekwurzel et al. (2017), and Licker et al. (2019)—make it possible to pinpoint Big Oil as a main agent bringing about climate change, the circuit linking oil and gas companies, fossil fuels, and the harm they cause must be closed in space, time, and scale so that they are reconnected in the consciousness of the wider public. This in turn entails climate science abandoning the manifest reductionism dictated by the hegemony of predictive natural sciences that could induce people to see climate change and the harm it causes as physically isolated processes independent from human agency (Hulme 2011). On the contrary, a more integrated and multidisciplinary approach to climate change would help broaden the evidence base and map out possible future scenarios and the ways for achieving them. In short, such an approach would make it possible to reconnect anthropogenic climate change to its root source, that is, the combustion of fossil fuels. In the end, only through the rewiring of our mindsets to make this connection can anti–fossil fuel norms develop and flourish globally.

In conclusion, it would be naive to expect Big Oil to change its behavior on its own. Only a strong societal focus on its responsibility and duties, backed and prompted by widely accredited anti–fossil fuel norms, can encourage it to accept—both through binding provisions and spontaneously—making reparations and transitioning to a cleaner business. As part III shows, this is by any means an arduous journey that involves resistance, struggles, and social, political, and economic risks. Hopefully these difficulties would not be comparable to those caused by the abolition of slavery: a bloody civil war in the United States that took the lives of more than 750,000 men (Hacker and McPherson 2011), by far the greatest toll of any war in American history. And for the British Empire, payments to slave owners totaled £20 ($27.0) million in 1834 (currently estimated at £23 [$31.1] billion), that is, 40 percent of the total government expenditure for that year, a sum borrowed by the British government that it only finished repaying in 2015 (Guthrie 2020).

SOCIETY AND BIG OIL'S DUTIES

Based largely on the harmfulness and moral wrongness of fossil fuels, society is coming around to exhorting Big Oil to abide by less harmful behavior by addressing its impacts and emissions debts: societal pressure requires fossil fuel companies to undertake predetermined actions to stop contributing to climate change. At the same time, there is a solid and composite moral basis for Big Oil's duties vis-à-vis climate change, grounded in the violation of the no-harm principle that compels oil and gas companies to follow operational behavioral guidelines. In other words, oil and gas companies must comply with certain duties. Such duties—morally grounded standards of behaviors—are not obscure philosophical concepts and should be considered as immediate and highly relevant informal *sanctions* to comply with what is required by society.

A sound justification of Big Oil's duties could provide a helpful framework for a reasoned dialogue with civil society as well as between political representatives belonging to different ends of the spectrum and subject to different constraints. Despite their alleged abstractedness, the duties of reparation and decarbonization are, in fact, societal-agreed moral provisions with major relevance to current international climate governance. These terms—*reparation* and *decarbonization*—reflect and emphasize the kind of actions required from the industry by society in light of its unique agency and moral responsibility.

Chapter 3 suggests that the duty of reparation can be understood as a specific and contingent form of the duty of adaptation, a form of *ex post* adaptation. In particular, the duty of reparation is intended as the financial rectification of climate-related harm, as explained in chapter 6 In other words, this duty implies rectification through financial means that is, through money, which should be disbursed by Big Oil to redress the suffering of people and communities due to climate change. In this perspective, the duty of reparation implies that people have a right to live in a world where they are not harmed by man-made climate change When this entitlement is not met, rectification is owed to remedy an unjust situation. On the other hand, when financial means are provided to maintain or restore people's ability to protect themselves from harm this counts as adaptation (Baatz 2017). According to this view, it seems straightforward to maintain that rectification with regard to the issue at hand concerns the right to not be harmed and therefore should be seen as a duty of reparation rather than within the perspective of the duty of adaptation. It is worth emphasizing that, as more exhaustively explained in chapter 9, in practical terms the duty of reparation is also meant to help displaced workers and frontline communities adapt, thereby increasing its political feasibility.

The actions demanded by the duty of decarbonization require, by and large, a reduction in carbon emissions to avoid/prevent future harm. This seems, as pointed out in chapter 3, to mirror the requirements of the duty of mitigation whereby, consistent with the UNFCCC (1992) and the IPCC (2014b), mitigation is understood as both the reduction of carbon emissions and the enhancement of carbon sinks. In fact, the duty of decarbonization is different and in many respects more stringent: it indeed requires Big Oil to mitigate its emissions. But more specifically, as already stressed, it means that oil and gas companies must reduce carbon emissions associated with their operations *and* products. However, the ultimate objective of the duty of decarbonization is that these companies phase out fossil fuels from their business in order to eventually eliminate carbon emissions. In turn, this would entail Big Oil ultimately changing its behavior by either ceasing operations completely or progressively transitioning to dealing in zero-carbon–intensive products, such as renewable energy, while of course keeping its operations carbon-free. In a nutshell, this duty requires Big Oil to eventually morph into Big Green Energy.

In sum, based on the demands emerging from society, the duties of reparation and decarbonization provide the moral basis for Big Oil's actions,
that is, for the possibility that oil and gas companies take action of their
own volition and indirect agents compel Big Oil to undertake practical
initiatives.

IS BIG OIL RESPONDING TO SOCIETY?

A few further questions on the relationship between the oil industry and
society in the context of the current climate crisis need to be addressed.
How is Big Oil responding to society? How much do oil and gas companies
accept the moral requirements posed by society and engage in activities for
their achievement? Obviously, Big Oil has not yet engaged in any reparation activities based on the theoretical provisions discussed thus far. Therefore, the attention of this section is inevitably limited to steps being taken
by the industry to reduce emissions, consistent with the requirements of
the duty of decarbonization. Dealing with the other side of the argument—
exploring how oil and gas companies actually operate in a climate-shaken
world—makes it possible to better understand their nuanced moral roles
and to more effectively shape their subsequent duties.

Big Oil, like other powerful corporate entities, has had a propensity
to shape environmental policy in more ways than simply supporting or
opposing regulations (Meckling 2015). As explained in chapter 2, until
a few years ago major US IOCs substantially adopted a reactive strategy based on the rebuttal of responsibility for climate change, whereas
European IOCs embraced a more proactive strategy that accepted some
forms of responsibility. For instance, Bloomberg's *Climate Transition
Score* highlights that European oil companies—especially TotalEnergies,
Portuguese Galp, Norwegian Equinor, BP, Shell, and Italian ENI—are
those most prepared for a low carbon-transition, while US oil companies
lag far behind (BloombergNEF 2021).

Despite a resurgence of a new cynical form of denialism that sows
doubts about the motives of those studying climate change and commu-
nicating their findings, things today appear to have changed, at least in

terms of attitudes and intentions. All the largest fossil fuel companies have recognized climate change and started to aim, with different objectives and paces, at a lower-carbon future.

For the first time in its history, the oil industry is consciously facing a new uncertain age in which climate change and the expansion of other low-carbon energy sources could downgrade its dominance and power. For instance, two leading IOCs seem to question their future business. BP, in its 2020 *Energy Outlook* (BP 2020b), affirms that oil saw its peak in 2019, due to the growth in renewable energy and consumers shifting to electric vehicles. ExxonMobil's 2019 *Energy Outlook* (ExxonMobil 2019b) acknowledges a similar peak in oil as well as in demand for gasoline, stressing that its assets may not be attractive investments in the near future. The Organization of the Petroleum Exporting Countries in its 2020 *World Oil Outlook* (OPEC 2020) instead claims that world oil demand will plateau in the late 2030s, and only then could it begin to decline.

At the same time, despite the intrinsic vagueness of net-zero targets (Rogelj et al. 2021) and the actual uncertainties associated with negative emissions technologies (Anderson and Peters 2016), Spanish IOC Repsol announced its plan to achieve net-zero emissions by 2050 and, importantly, stressed that its abatement commitment extends to scope 3 emissions, those originating from the downstream combustion of oil and gas Repsol has distributed within the global economic system (Storrow 2019).[1] Equinor vowed net-zero emissions from both its operations and products by 2050 (Coleman 2020). BP heralded an analogous ambition of net-zero emissions—at least in upstream production—in a press release in February 2020 (BP 2020b), although there were no claims that it would shelve plans to increase its oil and gas extraction. Shell followed suit in April 2020 and announced an ambitious strategy to become a "net zero energy business by 2050 or sooner" largely based on offset expansion via carbon capture, nature-based solutions, and the rapid growth of biofuels and hydrogen.[2] These targets, however, clash with its plans to invest $8 billion a year in oil and gas in the short term, compared to the $2 billion to $3 billion a year being tabled for nonfossil sources (Cooke, Sherrington, and Hope 2021). In fact, in an attempt to gain shareholders' endorsement, Shell called on them to vote on its climate and energy transition strategy (Ambrose 2021). Interestingly, Petrobras CEO Roberto Castello Branco dismisses his European peers' 2050 net-zero claims as pie in the sky: "That's like a fad, to make promises for 2050. It's like a magical year, . . . On this side of the Atlantic we have a different view of climate change" (Millard 2020).

At any rate, the same formidable industry that successfully denied anthropogenic climate change and slowed/halted climate policy for decades now seems to have not only recognized climate change but also acknowledged the inevitability of the low-carbon transition and the impact on its own business, as increasingly urged by its stakeholders. This is also consistent with the demands posed by larger investors to shift more effectively and rapidly toward a low-carbon future to sustain the global economy and increase the prosperity of their clients. With Big Oil being a highly capital-intensive industry that largely relies on outside investments, this plea is taken very seriously in oil and gas companies' boardrooms.

In the face of the enormous challenge the industry faces and fully aware that its recurrent pledges about a low-carbon future are to be taken with more than a grain of salt, major oil and gas companies' current carbon management practices should be examined, with emission-reduction pathways envisioned.

With regard to carbon management, it seems that by and large, almost all companies perceive climate change as a business risk—though to varying degrees—and are sensitive to the requests from governments, investors, nongovernmental organizations, and, more broadly, society (Sullivan and Gouldson 2017). As previously stated, European companies led the way in acknowledging climate change; non-European IOCs and national oil companies (NOCs) followed in their footsteps later and in more ambiguous and roundabout ways. For instance, Russia's Gazprom, while agreeing with the necessity to cut carbon emissions, envisions and frames decarbonization mostly within a political and ideological perspective (Nasiritousi 2017). In sum, Big Oil proclaims that it can come up with the necessary solutions to strike the right balance between reducing emissions and safeguarding economic growth and prosperity.

"Let's give up the climate change charade: Exxon won't change its stripes." This is the provocative title of an editorial published in May 2016 in *The Guardian* by Bill McKibben, the prominent climate change activist and journalist. It would be hard to argue with McKibben's assertion: despite an increasing portrayal of Big Oil as the villain in the context of climate change by many civil society actors and media outlets, the Texas giant seems impermeable to change; it even feigns a sense of climate urgency and rolls out a *climate charm* offensive based on bold green pledges, alluring ads, and virtue signaling its support for policy proposals against carbon emissions to hide its unwillingness to change (Mulvey et al. 2019).

But the truth behind the masquerade is that oil and gas companies have not actually changed their fossil fuel–centered behavior and do not appear to have any plans to do so significantly in the future, as the concluding chapter unequivocally shows. At the same time, civil society, in its broadest possible understanding, has started to mobilize, targeting these companies for their responsibility for climate change with a slowly snowballing demand that they rectify the injustices suffered by undeserving peoples due to the harm they generated.

The book so far has made a case for defining and justifying Big Oil's responsibility and duties for climate change and, consequently, for recognizing that oil and gas companies should become new central agents of climate ethics and policy. In other words, an extremely important group of agents in the context of climate change—oil and gas companies—must be repositioned in the climate narrative, from the most vilified *dramatis personae* to constructive agents of change. They should assume a role in global climate governance, proportionally befitting to the one they played in the climate crisis, along with states, individuals, and other agents.

Broadening the perspective of climate policy and politics from states to oil companies, on the one hand, opens up new possibilities for them to become part of the solution rather than, in the best-case scenario, passive bystanders who continue to profit from unremitting climate disruption or, in the worst-case scenario, actively obstructing any steps taken to remediate the situation in order to safeguard profits. On the other hand, the social condemnation of fossil fuels and the prospect of escaping the current carbon lock-ins are greatly privileged if oil and gas companies are recognized as primary moral agents in climate change with specific responsibility and consequent duties of reparation and decarbonization.

This chapter lays the groundwork for analyzing the operationalization and implementation of these duties in part III. In particular, the chapter investigates the theoretical/normative bases of such duties, delineates their main features, and presents their most critical social, political, and economic implications in light of the calls emerging from civil society. The chapter concludes by emphasizing how these duties are strictly interdependent.

CORRECTIVE JUSTICE AND DUTIES

As said, corrective justice, originating from harmful wrongdoing, helps focus on the harm produced by Big Oil and elaborates on the resulting actions required to rectify the injustice caused by such harm: corrective justice provides the theoretical/normative basis for justifying and outlining Big Oil's duties of reparation and decarbonization engendered by its moral responsibility. To articulate the corrective justice perspective in relation to these duties, the following must be identified:

1. The duty-bearer (i.e., the agent who should bear financial and other burdens of rectificatory actions);
2. The moral basis of the injustice (i.e., the moral principles that justify and define rectificatory actions to redress the injustice caused by the harm);
3. The structure of the duties (i.e., the specificities of the duties in relation to oil and gas companies);
4. The forms that rectificatory actions should take in relation to oil and gas companies (i.e., the specifications and concrete means through which rectification for the harm done should be attained);
5. The duty-recipients (i.e., the subjects entitled to rectification and the modality of the allocation of the rectificatory actions among them envisaged by the duties identified); and
6. The indirect agents of justice.

A thorough response to point 1 would be tautological, since this analysis obviously considers oil companies as duty-bearers and, more broadly, as collective moral agents in climate change. The rest of this section briefly addresses point 2, since it is common to the duties of both reparation and decarbonization. After a justification of the logic and the rationale of the, in many respects unprecedented, duty of reparation, the two following sections investigate points 3, 4, and 5 (i.e., the structure, the forms, and the recipients of the duties with regard, respectively, to the duties of reparation and decarbonization). Point 6 is not addressed in this chapter; the details of indirect agents will be more usefully analyzed from an empirical vantage point in part III, where, as previously mentioned, they are termed *agents of destabilization.*

The Moral Basis of the Injustice

Point 2 requires defining the moral principles that justify imposing the duties on the industry. Climate ethics literature usually refers in this regard to two backward-looking principles, the polluter pays principle

(PPP) and the beneficiary pays principle (BPP), and one forward-looking principle, the ability to pay principle (APP), already touched upon previously. The PPP distributes burdens—financial and otherwise—associated with rectificatory actions in proportion to past contributions to the individual agent's level of emissions. The BPP holds instead that proportionality in such distribution should be calculated based on the benefits that agents have derived from activities generating emissions. The APP posits that the quota of burdens should be proportional to the agents' relative capacity to bear such burdens.

All these principles aim to establish and justify a positive responsibility for sharing the burden of rectifying the unjust situation caused by climate change. Instead of relying on any single principle, the current analysis endorses a hybrid version (Shue 2015). Basically, this version holds that those who contributed heavily to creating the problem through their excessive emissions both benefited more and are better able to pay than most others. This *triply* hybrid account might be somewhat controversial with regard to some details of the principles included, but nonetheless they all converge at the practical core of reinforcing the moral basis of the duties of reparation and decarbonization. This view appears, in fact, to be perfectly suited to the oil industry and provides a moral justification for said duties, indeed with a different moral relevance of the principles included for these two specific duties.

For instance, the forward-looking cogency of the APP based on the agent's relative capacity to shoulder a burden as clarified by fact E, enrichment, might be more morally significant for the duty of reparation. The APP principle considerably strengthens this duty, as it better captures the *wealth component*—which in this case is very important given that the rectificatory action envisaged by this duty is, in fact, of a financial nature. At the same time, again in relation to the duty of reparation, the backward-looking BPP further justifies the moral cogency of fact E, enrichment. The PPP instead more forcefully backs the duty of decarbonization.

WHY A DUTY OF REPARATION FOR BIG OIL?

Before analyzing the structure, forms, recipients, and other relevant features of the duty of reparation (a similar analysis of the duty of decarbonization will follow), the logic and rationale justifying the urgency of reparation require explicit justification. It may in fact not be intuitively evident why Big Oil, besides refraining from causing further harm by decarbonizing its business, also has a duty to financially rectify the injustice caused to those subject to the harm caused by its activity.

The Logic

The mounting toll of climate change and the prospect of a further surge of threats in the future prompt a question that has so far been skillfully circumvented in the climate debate and negotiations: who should pay the costs? This is indeed an issue that has traversed the history of anthropogenic climate change, its international negotiations, and its science, at least since its acknowledgment in the early 1990s when the scientific evidence pointed to unevenly distributed harm and to the consequent necessity of fairly dealing with it. However, the enormity of the task of rectifying the harm done has substantially contributed to a paralysis in combating the climate crisis, as happened in other historical cases of patent wrongdoing such as slavery, Jim Crow laws, and survivors of the Holocaust demanding reparations from the German state.

Unfortunately, climate reparations are particularly controversial due to the fact that, as previously stated, the problems of climate change can overwhelm our cognitive and moral systems. Oil and gas companies, through their fossil fuel–related activities, produced widespread historical injustice in the form of climate-related harm; its rectification would produce a more just state of affairs. Therefore, given the satisfaction of both backward- and forward-looking considerations for reparation of past injustice, it seems sufficiently safe to claim that the oil industry has an obligation to rectify its wrongful actions that are resulting in negative climate impacts and harm if it is to retain its social license to operate.

This logic of climate reparations, despite a long history of neglect in climate negotiations, was eventually espoused by the 2007 United Nations Framework Convention on Climate Change Bali Action Plan, which suggested that richer countries assist particularly vulnerable ones in addressing the impacts of climate change. This proposal was operationalized through the 2013 Warsaw International Mechanism for Loss and Damage and established under the Cancun Adaptation Framework, which specifies that developed countries ought to assist particularly vulnerable developing ones in dealing with climate harm. The plan still suffers from a certain lack of progress toward its practical introduction, however. As for nonstate agents, though, the logic and objectives of climate reparations becomes an even thornier issue to tackle.

And yet Big Oil's manifest responsibility for the climate crisis has triggered various legal responses targeting climate liability, constitutional and human rights, and securities fraud. In other words, while waiting for anti–fossil fuel norms to emerge and for political authorities to convert them into binding provisions, tort law is a way, albeit imperfect, of achieving the goals dictated by ethics.

There is no lack of initiatives and political proposals, though. For instance, several substate plaintiffs in the United States are seeking to shift part of the cost of protection from climate impacts to fossil fuel companies, and California has enacted a ban on fracking-based oil drilling. As of September 2021, twenty-eight cases have been filed against major oil companies by cities, counties, and states in the United States.[1] Particularly significant in this regard are the words of New York City mayor Bill De Blasio in announcing his city's intention to sue BP, Chevron, ConocoPhillips, ExxonMobil, and Shell:

For decades, Big Oil ravaged the environment and Big Oil copied Big Tobacco. They used a classic cynical playbook. They denied and denied and denied that their product was lethal. Meanwhile they spent a lot of time hooking society on that lethal product; and think about how cynical and dangerous that is, knowing the damage that was being caused, having all the evidence in the world, and yet using all the tools at their disposal to deepen the crisis for their own profit. Were they punished for these destructive actions? No. They were rewarded to the tune of trillions of dollars. Well, today the nation's biggest city says no more. They won't be rewarded anymore. It's time for them to start paying for the damage they've done. (New York City Government 2018)

But that's not all. The lawsuit's introduction begins with this statement: "This lawsuit is based upon the fundamental principle that a corporation that makes a product causing severe harm when used exactly as intended should shoulder the costs of abating that harm" (Sabin Center for Climate Change Law 2021).

So far the legal strategy has yet to achieve its goals. For instance, the abovementioned New York City lawsuit was dismissed by a US District Court judge based on the argument that given its enormity, climate change must be addressed through federal regulation and foreign policy rather than through piecemeal litigation (Pierson 2018). Similarly, New York's attorney general failed to prove that ExxonMobil misled shareholders over the true cost of climate change (Stevens 2019). The words of the editor of *Sierra* magazine, Jason Mark, seem very opportune in this case: "The law is an imperfect extension of ethics. Tort law alone isn't going to save the planet. Even if, after years of litigation, the pending cases succeed, the question of climate restitution may well be too large for the courtroom, the damages too vast for any single judge or jury to decide" (Mark 2018, 12).

The Theoretical Rationale

A theoretical justification for the duty of reparation quite simply requires Big Oil to rectify the harm it has caused by providing various forms of support for those affected by it; in particular, rectification should consist of financial means. Exactly how is far from obvious, but the quantitative argument will be justified in the following section. Indeed, the general theoretical rationale to the duty of reparation is provided by the construct of positive moral responsibility pertaining to Big Oil, as defined in chapter 4. From a moral perspective, rectification of the harm done is a wholly legitimate claim by those who suffer it (Goodin 1989; Shue 1999; Miller

2008; Pogge 2009). A vast literature expressly targeting climate-related harm posits that those who are harmed by it are entitled to some form of rectification (e.g., Caney 2010; Shue 2011).

From a different perspective, oil and gas companies would be *voluntary beneficiaries* because they knew of their wrongdoings and could have avoided them without incurring unreasonable costs and instead intentionally pursued a path in full awareness of the consequences this would wreak (see the morally relevant facts of chapter 2). Thus, as voluntary beneficiary, the oil industry must rectify the harm done by supporting those affected by it. In terms of the moral principles involved, this theoretical view reflects and justifies the BPP, which holds that the burden of rectificatory actions should be calculated based on the benefits that agents have derived from activities generating emissions (Pasternak 2014). This is consistent with what was labeled in chapter 2 as the morally relevant fact E, enrichment. Oil and gas companies amassed extraordinary wealth and greatly increased the fortunes of their shareholders through their fossil fuel–related activities. As already stated, this is not in itself morally wrong. However, enrichment is still a morally relevant fact, one that strengthens and better typifies Big Oil's responsibility and duties for climate change. Enrichment responds to a moral logic that distributes the burden of rectificatory actions in proportion to the benefits derived and in a different perspective to the ability to pay too.

The Pragmatic Rationale

From a pragmatic perspective, Big Oil's wealth and power means the industry is actually able to financially rectify—at least—a substantial and morally relevant part of the harm done. This financial capacity must indeed take into account the trade-offs and limitations that the costly duty of decarbonization imposes on it. Nonetheless, if the duty of reparation is shaped in feasible and effective terms, Big Oil's wealth guarantees the ability to redress the harm done.

Usually the cost associated with climate change is borne by taxpayers as well as by affected individuals and businesses; one of the most urgent demands that climate and social justice movements make is, as said, that Big Oil repay its impacts debt (Mark 2018; Warlenius 2018). State-of-the-art science makes it possible to consider apportioning the responsibility for climate harm to individual oil and gas companies so as to meet the bottom-up demand for climate justice. In short, as discussed in chapter 9 the increasing capacity of attribution science is making it far easier to identify the relevant agents and therefore to fairly divide up the financial burdens required by the duty of reparation.

THE DUTY OF REPARATION: STRUCTURE, FORMS, AND DUTY-RECIPIENTS

The duty of reparation posits that oil and gas companies should relinquish a portion of the wealth deriving from their harmful activities in order to financially rectify the harm suffered by other agents.

To frame and better understand the structure of the duty of reparation of the corrective justice specifications reported above (point 3), the book considers these companies as moral agents that, through their harmful fossil fuel–related activities, have benefited from the suffering of others. Based on this violation of the no-harm principle, they have the (first-order) moral responsibility and the related duty to support affected agents.

There are different ways to support affected agents, ranging from immaterial approaches, such as public acknowledgment and apologies and the establishment of a truth commission (Rotberg and Thompson 2000), to material rectification of historical wrongdoing, vital in confronting harmful activities (Fraser 1995). In the context of climate change, much remains to be done in practical terms to reduce its harmful impacts; rectification therefore must be primarily material, mainly aimed at minimizing its impacts through practical actions.

There are different forms of material rectification, too. For example, *restitution* implies returning misappropriated things to the rightful owners or their heirs, while *compensation* means financially redressing the rightful owners or their heirs for the harm done (Goodin

2013). Unfortunately, since they both require identification of the duty-recipient, applying restitution and compensation forms of rectification is highly problematic, considering the complex nature of climate change. Given substantial temporal and spatial lags between carbon emissions and their impacts, it is virtually impossible to identify the rightful duty-recipient or a legitimate heir with certainty. Moreover, in the case of restitution, the context of climate change makes it close to impossible to identify the misappropriated thing (apart from a rather abstract notion of atmospheric absorptive capacity, which was wrongfully overconsumed by oil companies' emissions).

It should be noted that part of the relevant literature uses the term *compensation* as synonymous for what is meant here as *financial rectification*. This book instead considers compensation to be a specific form of financial rectification that requires the identification of the recipient of the funds. As said, financial rectification of the harm done is instead generally understood as a cash-based form of rectification. *Rectification* alone signifies a broader meaning that includes not only material forms (financial and nonfinancial) but also nonmaterial ones, such as recognition of blameworthiness, apologizing, and so on.

At any rate, where restitution and compensation fail, *disgorgement* appears to be more appropriate. Disgorgement is the relinquishment of profits or holdings acquired through past wrongful acts. It does not necessarily entail identifying the agents who suffered harm, who the legitimate heir in the current generation is, or how that person would have fared had the past wrong not occurred. Disgorgement demands only the relinquishment of the fruits of the historical wrongdoing: in the case of the oil industry, its tainted assets and benefits. Unlike the restitution and the even more demanding compensation forms of financial rectification, disgorgement focuses on the duty-bearer and not on the duty-recipients and their welfare (Goodin 2013). The objective of disgorgement is that financial rectification that cannot be provided in some way to the rightful duty-recipient can nonetheless be effectively channeled to a valid related cause: for example, in the case of slavery, to the cause of equal opportunities fo-

people of African descent, and in the case of climate change, to alleviate the suffering of the most vulnerable communities and stakeholders.

A remarkable example of implementing the provisions of disgorgement occurred in the case of the Nazi plundering of artworks from European Jews. After the war in instances where the victims of the theft were heirless, the art was sold, and the proceeds were put into a fund to provide support to Holocaust survivors. Similarly instructive is the example regarding Brown University's University Hall, which was built through timber and slave labor *donated* by wealthy benefactors. Since the slaves involved were unknown, it is impossible to compensate their heirs; disgorgement, however, provides an opportunity to financially rectify the wrongdoing associated with the building of University Hall by transferring some of the university's wealth to living people of one or more chosen groups (Goodin 2013). In the same vein, Glasgow University has committed to a reparative justice program to financially rectify its role in the slave trade (Baggini 2018). And the British pub chain and brewer Greene King and Lloyd's of London have pledged to pay financial rectifications to representatives of Black people and of other minority ethnic backgrounds to redress their founders' roles in the transatlantic slave trade (Rawlinson 2020).

Since disgorgement does not require the identification of a particular duty-recipient or speculation over how they would have been today had the past wrong not occurred, its potential and advantages lie in the informational parsimony that makes it much more feasible, especially in the complex context of climate change. Disgorgement, furthermore, would take into consideration only assets and benefits that are *tainted*, not those that are attributable to any oil and gas companies' actions. For example, tainted benefits would not include charity donations and benefits to communities that emerged as a result of oil-related operations. On the other hand, they should include all those benefits not employed in climate-productive ways, such as speculative financial investments.

Before analyzing the form of the duty of reparation, a specification is essential: from a moral perspective it is better to avoid harm in the first place by adapting to it than to rectify it, financially or not, in retrospect (Baatz 2018). In this regard, one argument seems very pertinent: in the event of irreplaceable loss, it is possible to distinguish between *end-displacing* financial rectification, which helps people in pursuing other ends that would leave them as well off as they would have been if the loss had not occurred, and *means-replacing* financial rectification, which provides people equivalent means for pursuing the same ends (Goodin 1989). The former is inferior because it obliges people to pursue other goals with other means. This distinction is helpful in considering financial rectification for climate harm. Therefore, adaptation is preferable to rectification, with the distinction being that the former corresponds to means-replacing financial rectification and the latter to ends-displacing financial rectification.

Having dutifully acknowledged the moral superiority of adaptation funding, nonetheless the sole objective of Big Oil's duty of reparation should be to financially rectify *ex post* the harm done. On the one hand, a narrower focus on harm rectification greatly simplifies the design and implementation of the duty of reparation's allocation structures and processes. On the other hand, as repeatedly underlined, the oil industry is only one of the moral agents with (first-order) responsibility and duties in the context of climate change: accordingly, adaptation funding should be one of the archetypical duties that political authorities adopt toward peoples threatened by climate change, given their general objective of protecting their citizens from harm. This is implicitly proclaimed by the United Nations Framework Convention on Climate Change when, in article 3.3, it posits that the parties (i.e., states) "should take precautionary measures to anticipate, prevent or minimize the causes of climate change and mitigate its adverse effects." Signatory states are duty-bound to prevent as much harm as possible rather than providing rectification *ex post*.

Big Oil's focus on rectifying the harm already done, moreover, reinforces the moral cogency of the duty of reparation, since, as specified above, disgorgement is in principle justifiable through scientifically agreed causative chains. Indeed, subnational political authorities that are seeking to hold oil and gas companies accountable for climate change (e.g., US municipalities such as New York City) focus on the cost of the damage (harm) generated by climate-related events like increasingly severe floods, storms, hurricanes, and sea level rising.

The ultimate goal of the duty of reparation is supporting the most vulnerable agents from the impacts of climate change by providing them with the adequate financial means for coping with the related harm and loss. This general consideration makes it possible to fully address point 5 of the specifications of the corrective justice perspective in relation to oil and gas companies' duties indicated above: who are the duty-recipients?

By and large, duty-recipients are the peoples and communities most vulnerable to climate harm as well as climate migrants and climate refugees. Vulnerability to climate change cannot merely be defined as exposure to certain harmful events; it is also about the preparedness and capacity to cope with the effects. In this light, it is useful to clarify the notion of vulnerability that, applied to social systems, is also termed *social vulnerability* (Brooks, Adger, and Kelly 2005). Social vulnerability could be broadly understood as a state of well-being pertaining directly to individuals and social groups. Its causes are related not only to climate impacts but also to social, institutional, and economic factors such as poverty, class, race, ethnicity, and gender (Paavola and Adger 2006).

Social vulnerability produced by climate harm involves a number of critical aspects of well-being, such as life, health, livelihood, and so on. In principle, the degree of social vulnerability can be used to define the duty-recipients' level of entitlement to the disgorged funds: the greater their social vulnerability, the larger the rectification through disgorged funds. The prominent philosopher Henry Shue's third general principle of equity clearly endorses a stringent normative imperative of putting the most socially vulnerable first (Shue 1999). Being socially vulnerable means having far less than enough, and the principle of a guaranteed minimum states that those who have less than enough for a decent human life should be given a sufficient amount to meet this minimal requirement. More socially vulnerable peoples and communities therefore should be given sufficient rectification means (the funds, in this case) necessary to cope with and recover from climate impacts.

At the same time, there is another group of vulnerable agents, perhaps not subject to actual climate harm but who could suffer a different kind of loss deriving from the shrinking financial capacity that the duty of reparation imposes to Big Oil (and indeed from commitments to the low-carbon transition required by decarbonization). These agents are the displaced workers of the industries—fossil fuel and other critical ones such as chemicals, transport, shipping, and fossil fuel industry suppliers—damaged in terms of job loss/reduction of opportunities by this transition as well as frontline communities along the fossil fuel supply chain: they can be defined as *direct victims* of a low-carbon transition (Sovacool 2021). It should be emphasized that the inclusion among duty-recipients of displaced workers and impacted communities enlarges the scope of the duty of reparation beyond the strict moral boundaries of the financial rectification of the harm generated by fossil fuel–related activities. The rationale for this choice is eminently pragmatic. On the one hand, a wider scope greatly increases the acceptability and feasibility of the duty of reparation. On the other hand, the establishment of a separate (indispensable) fund for displaced workers and impacted communities would probably be too cumbersome for the already overburdened international governance of climate change.

From a different perspective, not only should the duty of reparation strike the right balance with the duty of decarbonization, while bearing in mind the consequent gradualism and prudence required by trade-offs, but—largely, again for the sake of its feasibility—it should explicitly engage with the low-carbon transition. Therefore, part of the funds disgorged should be earmarked for this goal. This quota should address the requirements of the oil industry itself during the transition phase, of agents enabling the socioeconomic transition to a low-carbon economy (e.g., research and policy institutions, international organizations, and nongovernmental organizations), and of green technology providers.

THE DUTY OF DECARBONIZATION: STRUCTURE, FORMS, AND DUTY-RECIPIENTS

To address the harm produced by their fossil-fuel related activities, the duty of decarbonization compels oil and gas companies to engage in a process of eliminating carbon emissions from their business.

Decarbonizing the oil industry's business means adopting non–carbon-intensive business models to eliminate emissions from companies' operations and, above all, products. To truly decarbonize, an oil company would have to either cease operations completely or transition to distributing low/zero-carbon–intensive products, such as renewable energy; that is, it requires Big Oil to become the *Big Green* of the book's title.

This broad understanding of decarbonization should not be confused with two narrower interpretations. One would only compel oil and gas companies to comply with binding emission limits set by specific political and regulatory bodies (e.g., governments, environmental agencies, and local, national, regional, international authorities with enforcement power). This narrow commitment to decarbonize depends on the willingness of political authorities to set and enforce binding emission limits; the broader notion of decarbonization entails much thornier governance-related behavioral and institutional issues. Another limited interpretation implies only the decarbonization of the industry's operations, such as reducing the carbon footprint of offices around the world. Some companies have already engaged in such actions, which, in essence, have served the purpose of *greenwashing* their image. For instance, the same American Petroleum Institute that, as evinced in part I, contributed to denying climate change and to disabling climate policy spent an amazing $663 million on public relations and advertising the virtues of the products of its associates in the last ten years. This was a far higher figure than the $98.4 million that the combined US renewable energy trade groups disbursed with the same aim during the period (D'Angelo 2019).

At any rate, decarbonizing the fossil fuel industry's operations—and not its products—is clearly insufficient, considering that it distributes fossil fuels to the global economy and that the scope 3 emissions they produce are the lion's share (roughly 90 percent, as said) of the industry's emissions.

As underlined in chapter 4, the allocation of the burden required by the duty of decarbonization should, in principle, follow the indications of distributive justice. Since carbon emissions are the commonly accepted *currency* of climate ethics, framing and accounting for the burden of decarbonization imposed on Big Oil in terms of emissions is the logical course of action. In this light, decarbonization entails an extensive and systematic reduction in the carbon emissions generated by the products and the overall activities of the industry. In principle, a burden of this significance should be allocated between the companies in question based on the principle of historical responsibility for their cumulative emissions, a sound measurement of their harm-generating activity over time. The companies that most greatly contributed to cumulative global emissions should curb their fossil fuel–related activities faster than the less implicated ones. Any *carbon allowances* that may be assigned to individual companies according to this logic should be gradually reduced over time to zero. In practice, however, elaboration on ways to allocate the burden of decarbonization among the various companies can be done only with specific reference to an empirically feasible hypothesis and will be addressed in part III.

Given the global nature and spatial unpredictability of harm reduction generated by the decarbonization of the fossil fuel industry, humanity in its entirety is the duty-recipient, as are other specific categories, such as oil industry workers and oil-dependent communities mentioned above. Indeed, this perspective and, more generally, the analysis carried out in this book are blatantly anthropocentric: they delve no deeper than the level of human beings and intrahuman relationships, and all moral considerations are framed in anthropocentric terms. The intentional omission of nonanthropocentric moral systems does not imply any disregard for other perspectives. The reason for the anthropocentric stance of this work is due to its objectives: its ultimate aim is to set out a morally based reference for inducing Big Oil to meet the duties that its responsibility for climate change involves. In other words, the book aims to provide the groundwork for addressing—broadly understood—institutional/political/governance/economic issues, because the duty of decarbonization (and indeed the duty of reparation) consists primarily of institutional/political/governance/economic efforts embedded in socioeconomic systems. Hence, it is assumed here that despite the general moral controversies that anthropocentrism implies, non–human-centered moral paradigms are not necessary in this context of analysis. When dealing with the duties of reparation and decarbonization, anthropocentrism can appear more applicable and viable on a practical level.

BIG OIL DECARBONIZATION: TARGETING FOSSIL FUEL
PRODUCTION THROUGH SUPPLY-SIDE MEASURES

Considering that the combustion of fossil fuels is by far the largest human source of carbon emissions and that the vast majority of such fuels are operated by oil and gas companies, as shown in part I, such entities would be expected—or rather, as this book suggests, should be obliged or persuaded—to strongly limit their fossil fuel–related operations and products (i.e., to decarbonize their business). This is indeed an enormous and almost unprecedented challenge in the history of humanity, with impacts on socioeconomic systems similar in terms of relative scale only to those of the abolition of slavery in nineteenth-century slave-dependent economies such as the American South (Mouhot 2011; Hayes 2014; Jamieson 2017). Numerous thorny issues are inevitable.

The decarbonization of Big Oil should, in fact, be more conveniently accomplished through instruments and approaches of climate policy aimed at restricting the upstream supply of fossil fuels, as climate justice movements are also starting to demand (McAdam 2017; Green 2018b; Asheim et al. 2019), rather than merely through traditional ones that focus on the consumption side, such as emissions taxes, regulations, and measures to support demand for less carbon-intensive goods and services. The former measures aim to influence the pace and location of fossil fuel extraction and are referred to as *supply-side climate policy*, a relatively novel and yet underutilized approach to achieve climate goals (Lazarus and van Asselt 2018).

The major agents of supply-side policy are the oil and gas companies, so supply-side measures should apply primarily to them. As said, the duty of decarbonization requires these companies to commit to phase out fossil fuels by adopting non–carbon-intensive business models. This would entail either the drastic step of Big Oil having to cease operations completely or, in a manner that would be less disruptive to the global economy and demonstrating an amenability on the side of the industry, transition to low/zero–carbon-intensive products, such as renewable energy. To achieve these objectives, supply-side climate measures seem the most appropriate, despite their currently being somewhat overlooked in mainstream climate policy discourse. Indeed, it is worth recalling that effective climate policies should include both supply- and demand-side instruments; here the focus is only on the first, with the implicit assumption that they are meant to complement the latter; by targeting fossil fuel production, the duty of decarbonatization of oil and gas producers can better be shaped and operationalized.

By and large, supply-side climate measures aim at constraining and/ or influencing the production of fossil fuels—in terms of both rate and location—whose downstream consumption causes carbon emissions. According to the literature (e.g., Sinn 2008; Green 2018a, 2018b; Green and Dennis 2018; Lazarus and van Asselt 2018; Le Billon and Kristoffersen 2020), supply-side measures targeting the production of fossil fuels have the distinctive advantages of efficiency/effectiveness and political opportunity compared to those targeting consumption.

Economic advantages of supply-side measures include the slowing down of investments in infrastructures for oil and gas production and transportation, thus limiting the carbon lock-ins associated with fossil fuels discussed in chapter 10. Additionally, such measures help counter the risk that panicked resource owners preempt increasingly stringent future emissions policies by accelerating production in the near term, as the *green paradox* envisages. Finally, measures targeting fossil fuels have low administrative and transaction costs, higher abatement certainty (due to the relative ease of monitoring, reporting, and verification), and comprehensive within-sector coverage.

Political advantages lie mostly in the potential of supply-side measures to mobilize public support, favor international policy cooperation, and engage segments of the fossil fuel industry. More broadly, such measures can increase moral pressure on Big Oil, given that actions are more easily observable and identifiable and that consequences are relatively predictable. These political advantages are expected to favor the emergence of anti–fossil fuel norms.

Between 1988 and 2017, 1,302 supply-side measures had been implemented in 106 countries, with rapid growth in the past decade (Gaulin and Le Billon 2020). The most relevant supply-side measures for targeting Big Oil's fossil fuel production can be grouped—based on the original International Panel on Climate Change typology (Somanathan, Sterner, and Sugiyama 2014)—in economic instruments, regulatory approaches, and government provision of goods and services. With regard to the focus of this book, the most useful ones are those that aim at restricting the supply of fossil fuels: carbon pricing, subsidy reduction, production quotas, and supply ban/moratorium.

Carbon pricing—carbon taxes and cap-and-trade systems—are usually applied at the point of fossil fuel distribution or final consumption. Economic analyses (Metcalf and Weisbach 2009), however, suggest that a more efficient application would be to target upstream activities at the point of extraction for reasons of maximum coverage and minimum administrative costs (given an exceedingly limited number of producers compared to the masses of consumers). Global fossil fuel subsidies astonishingly amounted to roughly 6.5 percent of global GDP ($5.2 trillion) in 2017 (Coady et al. 2019), and they greatly increase the profitability of the oil and gas industry (Achakulwisut, Erickson, and Koplow 2021).[2] Their elimination could help contrast climate change by discouraging inefficient energy consumption and leveling the playing field for renewables.

Regulatory approaches aim at revising schemes for the extraction, production, and distribution of fossil fuels in several ways, ranging from restricting the use of state-owned land and waters to ceasing the development of resources or infrastructures. The most effective seem to be fossil fuel bans and the prohibition of the production of fossil fuels and of further construction of related infrastructure. Economically inefficient though they may be, bans send the explicit signal that certain fossil fuel–related practices are no longer acceptable; therefore the logic of bans resonates with the claims of anti–fossil fuel movements and favors the emergence and diffusion of anti–fossil fuel norms.

Further steps to take could include governments acquiring production rights and finding ways to compensate the industry for not developing fossil fuel reserves or for restricting production. Governments can also restrict funding to the fossil fuel industry by divesting state pension funds and investment funds from the companies involved in such products and can encourage multilateral finance intuitions to withdraw from fossil fuel investments or support social movements that promote fossil-free financial options.

THE DUTIES OF REPARATION AND DECARBONIZATION: INTERCONNECTIONS, TRADE-OFFS, AND MAJOR CHALLENGES

Reparation and decarbonization are strictly intertwined, with trade-offs between them. At the same time, such duties—the duty of reparation in particular—face major challenges at multiple levels. To prepare the terrain for part III to more effectively sink its teeth into the operationalization and implementation of the duties, this section tackles this issue.

Interconnection and Trade-Offs

Reparation of the harm done by anthropogenic climate change drains substantial resources for decarbonization; by the same token, decarbonization is costly in terms of financial resources and it can indeed leave little scope to pay for reparations. In short, both duties are financially demanding and are expected to crowd out one another. An ideal balance between the two must be achieved: given the interconnectedness and trade-offs between Big Oil's duties, there may be different pathways for them to evolve.

In principle, the harshest (and least likely) possibility would involve abrupt dissolution of oil and gas companies as a result of the immediate termination of their fossil fuel–related activities. This abrupt termination would be the most effective option to prevent future harm. However, given the trade-offs noted above, it would at the same time deprive victims of climate change from fair rectification for the past harm suffered. This scenario would also jeopardize some of Big Oil's more vulnerable shareholders, such as pension funds and their individual investors. So, although attractive from the perspective of preventing future harm, this scenario is not functional from the point of view of rectification of past harm (i.e., of the duty of reparation).

Financial rectifications for more vulnerable stakeholders make a strong case for keeping oil and gas companies functioning and profitable so as to enable them to serve up justice where required. The option of phasing out fossil fuels from operations and products more gradually would certainly be less disruptive than a wholesale industry shutdown for the fossil fuel–dependent global socioeconomic system, including the interests of some countries (especially in the case of national oil companies [NOCs] and of oil-dependent countries) and other businesses that rely on fossil fuels (such as the chemical and automotive industries). This, however, does not alter the ultimate goal of decarbonization, which entails a complete phasing out of fossil fuels by the industry over a period of several decades.

This transition can assume various forms in terms of time spans and proportional combinations of reparation, decarbonization, business as usual, etc. The possible transition scenarios range—at the least desirable end of the spectrum—from slow and ineffective business as usual coupled with greenwashing efforts and business as usual coupled with enhanced reparation through financial rectification efforts to a more rapid phasing out of fossil fuels combined with a switch to other non-carbon-intensive business models and proportionally less commitment to reparation.

In the end, trade-offs between the duties of reparation and decarbonization are inevitable. Big Oil, though enormously wealthy, has a finite budget to allocate to the two duties and would need to prioritize the more morally appropriate courses of action. In many respects, it would be possible to posit that from a moral perspective, the industry's capacity to fulfill its duty of reparation—while not jeopardizing its weakest stakeholders and potentially disrupting the global financial system, as chapter 8 evinces—is the main reason for not invoking its abrupt dissolution.

It would, however, be opportune to specify that at some point in the future Big Oil should have a duty of *full* decarbonization requiring it to completely clean its business, allowing it to morph into Big Green. Basically, oil and gas companies would have to phase out fossil fuels gradually; a reasonable timeline for phasing in clean energy could be set within the next two or three decades. In this way, they would not undermine rectification efforts or the legitimate interests of countries (in the case of NOCs), shareholders (such as pension funds), other businesses (e.g., those relying largely on fossil fuels such as the chemistry and automotive industries), individuals who cannot quickly switch to non–fossil fuel–related appliances, and, more generally, all other agents who are locked into current carbon-intensive socioeconomic systems. Society could reap more benefit by allowing Big Oil to maintain its capacity to generate resources to undertake meaningful decarbonization.

A further complexity arises from the circumstances evinced in part I: oil and gas companies, considered individually, lack consistency; they are a startlingly inhomogeneous group. For instance, Big Oil is usually divided into privately owned international oil companies (IOCs) and state-owned NOCs, characterized by remarkably different strategies and objectives that inevitably entail different accounts/levels of responsibilities.

Accordingly, the scope and depth of duties of reparation and decarbonization vary from one company to another. A thorough analysis of the implications of this issue is given in part III, but here it is worth briefly clarifying some distinctions with a couple of significant examples. One can, for instance, be ascribed to the companies' varied responsibility, which requires different degrees of emissions abatement depending on the specific cumulative emissions. Or, with regard to denial, it was only a handful of IOCs—admittedly with the not impartial acquiescence of the rest of the oil and gas companies (including NOCs)—that conceived and deployed the denial and opposition campaigns. This would imply that the IOCs most involved in denial should bear greater burdens; the same logic also applies to those IOCs with greater awareness of the perils of fossil fuels.

Yet, it remains a challenge to endorse one duty over the other in abstract terms: both reparation and decarbonization are critical from the moral perspective. As alluded to in the introduction, part III will intensify its gaze on the issue of how to contextualize this conundrum, offering a more nuanced exploration of the relative burden of both duties in relation to the twenty biggest oil majors, personalized, so to speak, into three distinct levels of requirements that take into account social, institutional, economic, political, and operational contexts. Chapter 9 includes a breakdown of factors that justify the magnitude of contributions to a suggested fund to be created to financially rectify harm caused, as demanded by the duty of reparation. Chapter 10 provides a detailed road map for Big Oil to progressively abandon carbon-intensive business models as posited by the duty of decarbonization.

Major Overarching Challenges

As said, IOCs are private entities whose business operations generally cover the full cycle from exploration through production and refinement to distribution of petroleum products. NOCs are by and large similarly structured, but they are fully or largely owned by a government. These differences must be kept in mind—although not to the exclusion of a general inclusive picture—when analyzing the significant problems in implementing the duties of reparation and decarbonization by Big Oil First and foremost, one of the stumbling blocks likely to be encountered is the general recognition and self-perception by oil and gas companies as the corporate entities responsible for the climate crisis. *Commonsense morality*—so to speak—would suggest that other industries (e.g., automotive, chemical, or construction industries) are similarly responsible for climate change, as they also continued and promoted the use of fossil fuels after the consensus on their harmfulness was established.

To address this challenge, it is essential to further emphasize the unique role of Big Oil in the current global socioeconomic system: these companies are the corporate entities that have been dictating the rules of the game in terms of reliance on fossil fuels to other businesses. Through their informed choice to continue the extraction, refinement, and distribution of fossil fuels in the 1990s, Big Oil perpetuated the dependency of other industries on their products, industries that had to shape their business models around fossil fuels. Therefore, oil and gas companies should be considered the main duty-bearers. Other industries that depend on supply from oil companies should be attributed fossil fuel–related duties only after the *rule-of-the-game shapers* (i.e., oil and gas companies) have met theirs. Identifying Big Oil as a stand-alone group, with very precise and unique moral responsibility, is crucial to advancing efforts to combat climate change.

The disruptive consequences of deeming oil and gas companies as agents of justice responsible for their actions poses a further major challenge. This issue lies fundamentally in the novelty of the problem. States have been the main agents of action against climate change for decades. Holding private and semiprivate agents (i.e., IOCs) accountable for their harmful activity usually falls within the jurisdiction of national and international courts. Recognizing the oil industry as morally responsible for climate change—as a group and as individual entities—with clear responsibility-bearing and subsequent duties would set a precedent and disrupt the status quo of the international system. There is no existing institutional structure that could accommodate the new arrange-ment and facilitate their assumption of responsibility. For instance, creat-ing a new global structure to collect and manage disgorged funds would raise challenges about ensuring its justice and legitimacy, the mode of participation, and the extent of private agents' obligations and rights. Yet this seems the only viable solution. Moreover, the current state-centered system that imposes constraints and conditions onto business entities would also challenge the dominant paradigms on the role of the state. Even though climate change puts an end to business as usual, there is likely to be resistance in part of the Western world to the idea of an inci-sive role of the state in the economy that conditions the operation of pow-erful global corporations.

This introduces another critical challenge, a motivational problem, so to speak. It is not hard to imagine that for reasons of self-interest, many influential shareholders and board members would be more inclined to press for a business-as-usual approach, or a greenwashed one, in control-ling the activities and future of these companies. This behavior should be condemned on moral grounds, since it prioritizes the wealth and power of the few over the lives, health, and wealth of many. However, it is not always a clear-cut case of greed versus virtue and vulnerability. The blurred lines between private and national interests and ownership structures—that is, IOCs versus NOCs—in many oil companies com-plicates the matter, as fossil fuel exports strongly affect the economic development of several natural resource–dependent states, such as Rus-sia, Saudi Arabia, and Venezuela. Resistance to any attempts to dissolve a corporation that is the primary source of economic growth and of fiscal revenue is inevitable, as such actions would directly endanger the econo-mies of these states.

These complexities and others are addressed in part III, which will ana-lyze how to operationalize and implement the duties of reparation and decarbonization in relation to the top twenty oil and gas companies, bear-ing in mind the political, social, and economic contexts they operate in.

III WHAT BIG OIL MUST DO

In April 2021, Spain approved a road map to carbon neutrality by 2050 that includes targets to reduce greenhouse gas emissions and make its energy consumption 74 percent renewable by 2030, far beyond the 32 percent target defined by the European Commission for the same year. At the same time, Poland remains the only member of the European Union not formally committed to the union's midcentury climate neutrality goals set by the Green Deal, despite pressures from the European Union itself and widespread criticism from people living in the Polish smog-blanketed coal regions.

Climate politics still seems to zigzag between goal posts and dead ends like an erratic pinball, setting off alarms or celebratory jingles depending on points of view, despite the tectonic shifts it has undergone in the past few years. The Paris Agreement—after years of frustrating paralysis—settled on a bottom-up system, allowing countries the flexibility to periodically set their own goals and ratchet them up over time. This approach basically transfers the bulk of the politics of climate change to the domestic level, where it both gains new impetus and exacerbates divergences, especially over the enormous social and economic implications of addressing the climate crisis.

Until a few years ago, climate politics was chiefly rooted in distributional issues; the debate raged, for instance, on who should pay more for energy after price hikes to disincentivize the use of fossil fuels. Now, amid the global environmental crisis, climate impacts have become altogether more disruptive, modifying the challenges, creating new obstacles, and upping the stakes to the *nth* degree. Merely agreeing on how to allocate resources seems like a throwback to a bygone era now that the issue has been recast in an existential vein, with the debate centering on whose model of life can survive (Colgan, Green, and Hale 2020).

On the one hand, if the climate crisis degenerates too greatly, there are vulnerable peoples (e.g., residents of small archipelagos in the developing world and coastal areas, climate migrants and refugees), communities, workers, and businesses (e.g., farmers, the tourism industry, other businesses along the fossil fuel value chain) who risk losing everything. On the other hand, the industry that most contributes to climate change risks going out of business if, in the event of fossil fuels actually being phased out, it does not undertake costly and problematic processes of decarbonization. And that is without even considering the masses of people employed by the industry losing their livelihoods. Both groups are expected to engage forcefully and pour vast resources into aggressively defending their self-interests. For instance, while civil society increases the pressure to break free from the current unsustainable fossil fuel–based model of development, Big Oil might respond with more short-termism and obstructionism on climate policy.

So, this new *existential* nature of climate politics is inevitably leading to more intense clashes. However, it also creates an opportunity to boost support for climate action through the formation of (ever-changing) coalitions of climate *winners* and *losers* within economic, social, and political dynamics. Pragmatically, with a view to acting on climate change, politics here is intended to mean getting people with conflicting ideas and objectives *to act alike, not simply to think alike*, to borrow a well-worn phrase from the influential US political commentator Walter Lippmann.

Within this landscape of increasing uncertainty and hostility, this chapter aims to examine the intersections between oil and climate politics and their implications for the oil industry. The ultimate objective is to chart a course for an as yet unexplored path, one that sees Big Oil putting into operation and implementing its duties of reparation and decarbonization. The top twenty oil and gas companies will be bundled into homogenous groups according to dynamics of power and governance dictated by political, social, and economic determinants and by objective parameters that directly influence the achievement of their duties of reparation and decarbonization, with the aim of identifying groups of companies with comparable levels of requirements demanded of them.

THE OIL COMPLEX AND BIG OIL IN A CLIMATE-SHAKEN WORLD

This section analyzes how the intersections between climate and oil politics and their political economies frame the context the industry operates in and thus paves the way for an investigation into how the companies can realistically meet their duties of reparation and decarbonization.

The landscape of the oil world is a vast and complicated one: the key actors are governments, subnational political authorities, the oil and gas industry, other industries, governmental and nongovernmental organizations, economic and financial institutions, and civil society more generally (Mitchell 2013). This world is developed around a global production network that intersects extensively with the political and economic interests of the wider world. Governments and oil and gas companies—the latter by exploring, producing, refining, and distributing their product—are the big cats dominating this *oil complex* (Watts 2005), with large vertically integrated, multilocation international oil companies (IOCs) and national oil companies (NOCs) at the core. The strategies and relationships of the latter both with each other and with governments as well as the politics and political economy of climate change are particularly critical at this moment in time.

Except where drilling occurs in the United States on nonfederal lands oil and gas formally belong to states. Tensions between oil importers and exporters and between states over supplies, competition among oil exporting countries, and conflicts between resource seeking companies and reserve holding states make oil politics almost intractable. Yet the oil complex has recently come under attack from new agents who advance social and moral issues such as climate justice, human rights, a just energy transition, poverty reduction, and climate-neutral finance and question the overall role and responsibility of Big Oil. Climate change undermines existing socioeconomic systems and their foundations at different levels; first and foremost is the political ground.

So, it would be opportune to understand how Big Oil implements its most relevant ideas, conceptualizes resources, and deals with power structures and institutions, configuring their relationship in the current climate-shaken world. The connection with technology in the broader political economic, and ecological context is also put under the microscope.

The first point that should be stressed is that the dominant *carboniferous* model of socioeconomic growth has created powerful entrenched interests that resist change. This holds true for Big Oil, for countries that depend heavily on fossil fuel internally, and for those that base their wealth on its export, the so-called *petrostates* (i.e., oil- and gas-exporting states that have been unable or unwilling to diversify from oil and therefore rely on oil and gas exports as their main income and have thus by and large opposed progress in domestic and international climate policy). Oil companies, in particular IOCs, have vigorously resisted climate policy and undermined climate action for decades especially in the United States, with obvious global consequences. For instance, a panel of United Nations Framework Convention on Climate Change and International Panel on Climate Change experts considers opposition from them as the main obstacle to achieving the 2°C objective (Kornek et al. 2020).

Big Oil's huge investments can be amortized only over long periods of time, so the companies fight tooth and claw to protect their vested interests. Left to their own devices, they would remain operational to generate high emissions for decades, resisting change, thanks to the strong connections with political parties, politicians, and state bureaucracies. To some extent there seems to be a correlation between the interests of governments and those of Big Oil, thus cementing the oil complex and limiting serious alternatives engendered by external shocks (Phelan, Henderson-Sellers, and Taplin 2013).

In the discourse and negotiations on climate change, a basic assumption—one almost universally agreed upon—is that fossil fuel–derived energy is the main driver of economic growth. The *untouchability* of Big Oil is mainly based on this key assumption that also largely explains how the oil industry's lobbying efforts have been so efficacious: the interests of states in ensuring that climate change governance does not create any obstacles to economic growth coincide with the interests of industry giants (Newell and Paterson 1998; Levy and Egan 2003). It is likely that the question of economic growth will continue to dominate climate policy and politics, despite the ample evidence that there are some major sectors—health care, insurance, tourism—that would greatly benefit from decarbonization. This different narrative would, however, need a heavy-duty supporting coalition outside of those sectors as well as civil society agents backing decarbonization.

The bottom line is that within the oil complex, fossil fuel companies, governments, industry representatives, institutions, and international managerial elites form a seemingly impenetrable barricade of interests—which the Italian Marxist philosopher Antonio Gramsci famously christened a *transnational historical bloc* (Gramsci 1975)—able to exercise instrumental, discursive, institutional, and material power to ensure that approved policies would not undermine the centrality of fossil fuels (Levy and Newell 2002). Indeed, power is central in the dynamics of Big Oil in relation to climate governance. Power, for this book's intents and purposes, is understood as the capacity to mobilize resources to resist fundamental system alteration (Avelino and Rotmans 2009) under the circumstances highlighted by a political economy approach (Levy and Newell 2002; Phelan et al. 2013).

At least four distinct issues are crucial for understanding Big Oil and its role as part of a hegemonic bloc from this standpoint: coalition formation in climate policy, the endeavor for legitimation, external attacks on the industry, and the role of governmentality.

First, the highlighted perspective of political economy frames the process of coalition formation through the Gramscian notion of hegemony according to which a transnational historical bloc tries to coalesce subordinate forces (e.g., labor, civil society, education, media, business) around specific ideologies. Hegemony includes an ideational dimension, which provides the ideological *glue* to keep coalitions together; in Big Oil's case hegemony provides the monopoly of the discursive power that legitimizes it and provides it with a social license to operate (Blondeel 2019). This makes it possible to understand the tensions between Big Oil's interests and corporate interests more favorable to climate action (e.g., the insurance sector) and the struggles between business agents and social movements advancing more aggressive action on climate change.

On the one hand, therefore, is Big Oil's regime of accumulation, which aims at articulating and steering labor regimes, production technologies, and consumption patterns to generate a clear and relatively smooth pattern of fossil fuel–based growth; Big Oil thereby avoids decarbonization through the deterrence of the energy-growth ideology (Levy and Egan 2003) and by employing the illicit means analyzed in part I. Using the same leverage, Big Oil could also escape any claims for rectification for the harm caused by the combustion of its fossil fuels distributed throughout the global economy. On the other hand, the notion of hegemony suggests that a bloc of interests to drive reparation and decarbonization also needs to build a coalition of forces with adequate political clout.

Second, political economy insights underline how oil companies are caught in the middle of the traditional logic of accumulation and a need for new forms of legitimation to address the demands by consumers, governmental agencies, their own employees, and shareholders to respond to climate change (Paterson 2010). This latter dynamic is at times simplistically dismissed as greenwashing, but as it often responds to complex forces and narratives for reconstituting the identity of firms, greenwashing can appear facile and reductive in some instances (Wright and Nyberg 2014).

Third, the notion of hegemony allows for a better framing of society's response to Big Oil: social movements in particular act in counterhegemonic ways by mounting critiques and opposition to the industry, as chapter 8 shows. Despite a history stretching back many years, the activity of social movements in climate change started in earnest in 2009 around the time of the fifteenth Conference of the Parties (known as COP 15) to the United Nations Framework Convention on Climate Change, held in Copenhagen, December 7–18 (Hadden 2014). Social movements' discourse originated and still coalesces around two main issues: a demand for climate justice and the refusal of the commodification of the climate crisis through carbon markets. Anti–carbon market movements emerged in the run-up to the 1997 Kyoto Protocol. Their main argument was that carbon markets were simply another tile in the neoliberal mosaic that, by allowing richer agents to buy their way out of emission abatements, were aimed at establishing new forms of environmental discrimination and colonialism.

This cultural and sociopolitical milieu gave birth to the climate justice movements, which explicitly saw climate governance in terms of political economy, thus broadening the frame of reference for contesting the current state of affairs. In this vein, they developed and operated strategies expressly opposing Big Oil, such as the "*Keep It in the Ground*" and *divestment* initiatives. Climate movements targeting Big Oil by and large had—and still do have—the objective of advocating democratic control over the economy. To this end, they recommend that climate challenges be governed via participatory practices rather than the technocratic ones that serve Big Oil's interests so well and acknowledge both the subversion that decarbonization would cause to the oil complex and the injustices produced by both climate change and the industry itself.

Fourth, the issue of political economy refers to the pervasive use in climate change management—especially by Big Oil—of approaches of *governmentality*, that is, of actions, discourses, and other initiatives that aim primarily at controlling and shaping people's attitudes in order to modify their behavior and expectations, thus bringing them in line with the interests of the oil complex. Big Oil, through the funding and orchestration of denial campaigns, has succeeded in modifying people's practices and identity.

For instance, through the narrative of *green capitalism*, Big Oil has depicted corporations and markets as the only means for responding to the climate crisis. Accordingly, several oil and gas companies, with the acquiescence of certain accommodating media outlets, have promoted an image of sustainability that pledges no conflicts or trade-offs, claiming that they can adequately address climate change while expanding global production of fossil fuels and shifting the blame squarely onto the shoulders of their very own consumers for their lack of green virtues, as shown in chapter 5. In truth, these are smoke-and-mirror sleights of hand aimed at thwarting action through a masterful control of the oil/climate nexus governmentality. This suggests that climate-related actions to contrast Big Oil should primarily win the hearts and minds of the greater public if they want to successfully uproot the embedded sense of uncertainty toward scientific facts.

BIG OIL'S HEGEMONY IN CLIMATE GOVERNANCE

Since the 1980s—starting from the Western world, especially from the United States and the United Kingdom, and gradually going global—a distinctive model of economic growth fundamentally changed how wealth, business, and work function and are acknowledged in society. It is by and large based on privatization, deregulation, lower taxes for businesses and the wealthy, more power for employers and shareholders, and less power for workers and other stakeholders. Immense efforts have been made to present it as inevitable and to depict any alternative as impossible.

The current system of climate governance has emerged within this ideological context. This system is a relatively loose arrangement involving significant contestation as well as collaboration among states, corporations, nongovernmental organizations, and multilateral institutions. Within this system, states act mostly as agents concerned with their own economic interests and their worldwide competitive edge; corporations, on the other hand, besides being the primary economic actors, are important political agents with significant policy influence. Market-based instruments (e.g., emissions trading, carbon pricing) are widely believed to be the most effective tools for addressing the climate crisis. These features have shaped a fragmented and flexible climate governance, which on the one hand could quite easily evolve and on the other hand embeds a fundamental source of weakness.

The oil complex, in particular oil and gas companies, are major agents of climate governance. It has been provocatively argued that "anyone who ignores that basic political economy [of the oil complex], who believes oil and gas companies will be good-faith partners in a climate-emergency effort, is indulging in a kind of willful naivete that is entirely too common in the carbon wonk community" (Roberts 2019a). It is necessary therefore to thoroughly investigate their role in a system of marketized climate governance in terms of both relationships with states and other sources of power and within the industry itself.

Given the insightfulness of the political economy approach in relation to Big Oil, it seems particularly useful to understand this role through the framework for analyzing business strategy in environmental governance provided by Levy and Newell (2002, 2005) and Levy and Egan (2003). By bridging macro- and microlevels of analysis, this approach suggests that governance is the outcome of a process of bargaining, compromise, and alliance formation between several agents such as states and transnational organizations, businesses and their associations, and social forces such as environmental groups and unions. Each governance system is shaped by microprocesses of bargaining and enveloped by macrostructures of material capabilities and ideological formations. These macrostructures constrain the bargaining processes within regimes through composite and fluctuating preexisting relations of power.

The Gramscian framework proposed in relation to climate politics and governance suggests a strategic concept of power that emphasizes the role of business agents, indicates civil society as a battleground of political struggle, and provides space for contestation by different groups in complex dynamic social systems. This framework can therefore facilitate the understanding of processes of power and conflict at the governance system level as well as the relationship between such systems of governance and the broader sociopolitical context.

In this perspective, the Gramscian approach reflects the negotiated nature of governance: even the most powerful oil major is unable to directly determine government policies or write the rulebook of climate governance. This is not, however, a matter of equal bargaining power among interest groups; Big Oil has substantial instrumental, discursive, institutional, and material power—as explained below—to hold sway over decision makers (Levy and Newell 2002).

This situation testifies to the formation of a historical bloc around fossil fuels aligned with Big Oil's interests. For a stable system of climate governance to emerge, major agents of the oil complex must also share common frames of understanding of the main features of climate change science and of the policy approaches to address the issue. Bargaining over governance system structures and processes engages the industry in strategic economic, technological, and political moves across its multiple bases of power. Such bargaining blurs the distinction between political and market strategies: any threat to the industry's markets is both an economic threat and a challenge to hegemonic stability. Subsequently, Big Oil's responses to such threats are both economic and political. The emerging governance system reflects this bargaining process in terms of economic, organizational, and discursive structures that align the interests of major agents and tends to be relatively stable.

In this milieu, the oil industry negotiates—it forms alliances, it compromises—so as to construct a hegemonic coalition with governmental agencies, nongovernmental organizations, segments of civil society, the media, and the epistemic communities with the capacity to shape and promote policies, norms, and institutions to structure the climate governance system in specific ways. It is no coincidence that one word keeps cropping up in this chapter: *hegemony*. In the Gramscian understanding, this term emphasizes the interaction of material, discursive, and organizational practices, structures, and tactics to sustain Big Oil's dominance and its legitimacy even in the face of the climate crisis. These companies adopt strategies to improve their market position and technological prowess, strengthen their legitimacy, discipline the labor market, and influence government policy (Levy and Newell 2002). Additionally, Big Oil founded and funds transnational industry groupings such as the Global Climate Coalition and the Climate Council, as clarified in part I of the book.

In brief, hegemony provides the basis of power. It thus seems reasonable to claim that Big Oil, to maintain and boost its position of power in climate governance, has engaged in a *war of position* across each of the pillars of hegemony. On a material level, the industry has developed strategies to secure existing and future market positions by, for instance, establishing international partnerships between IOCs—which still have the most sophisticated technology for producing *difficult oil*—and market-seeking, oil-rich NOCs.

On the discursive level, with the support of certain factions of an obliging platform-offering media, these companies denied that climate change was happening and sowed doubts about the veracity of the science attributing it to anthropogenic causes; they also set out to manipulate the public mindset by portraying themselves and their products as *green* through a language of sustainability, stewardship, and corporate responsibility, virtue signaling their corporate backing of a carbon tax driven by an industry-friendly logic.

On the organizational level, industry powerhouses constructed multifaceted coalitions across sectoral and political boundaries, able to penetrate deep into the heart of civil society and politics. There are countless examples of such practices. For instance, Big Oil sponsors culture (such as the highly controversial BP sponsorship of the British Museum and

London's Science Museum accepting funding from BP, Equinor, and Shell) and education (e.g., energy and climate research centers in major US universities). Similarly, it is openly acknowledged that the oil industry has long bankrolled sympathetic political parties and candidates (Goldberg, Marlon, Wang et al. 2020), buying its way around accountability in a number of countries worldwide. The oil majors have been systematically adept at securing their interests in ongoing political negotiations and policy actions related to climate change. First, they routinely establish relational networks with senior policy makers who provide policy access to the company. Second, given such contacts, a subtler mechanism of influence comes into play, with policy makers internalizing company's ideas and interests; thus, the voices of the business and political elites are often in unison in defining problems and desirable solutions. Third, Big Oil is strategic about its use of facts and information. It applies organized pressure, lobbies relentlessly, and throws opens its coffers to financial incentives and other measures to influence policy makers.

It is worth mentioning that conflict within the oil industry itself has also played a role in the oil complex and in shaping its relationship with climate governance systems. In a neopluralist perspective—one in which the expertise and resources of corporations make them important players, without attributing them structural power—the oil and gas corporations are not a monolithic group; on the contrary, they experience severe conflicts arising from the differential effects that varied climate measures have on individual companies. On the one hand, these intercompany conflicts often limit overall business power. On the other hand—and more importantly in this context—such conflicts may improve the capacity of civil society at large to exert pressure on companies and effect change in corporate behavior. In the event of potential clashes, such as between companies more or less willing to engage with renewables or between companies competing for the same resources, social movements and activist groups can exploit these divisions and form political alliances with companies whose positions most closely match their own.

IOCS AND NOCS: DETERMINANTS FOR REPARATION
AND DECARBONIZATION

Big Oil is not a homogenous group, as already stressed: for the sake of the current argument, privately owned IOCs are substantially different from publicly owned NOCs in terms, at least, of sociopolitical context, strategies, and objectives. In order to frame IOCs' and NOCs' actions in relation to the duties of reparation and decarbonization, the relevant features of such duties must be examined in light of the dynamics of power and politics delineated in the analysis carried out so far. The resulting categorization helps distinguish between different groups of oil and gas companies in view of operationalizing and implementing their duties of reparation and decarbonization.

Be forewarned, however, that this book does not overstate the traditional IOC/NOC dichotomy, as both groups, albeit in different ways and on different temporal scales, have contributed to the climate crisis and share, to varying degrees, the ensuing responsibility and duties. To operationalize and implement the duties of reparation and decarbonization, it is of greater use to group the companies according to other features that cut across the IOC/NOC divide.

It is initially worth recalling that IOCs are privately owned companies based in Europe and the United States whose business operations traditionally cover the full oil and gas cycle worldwide. NOCs by and large have a similar structure, but they are wholly or largely owned by a state government and are generally based in non-Western countries, both oil-rich and not, apart from Norwegian Equinor, which is not, however, among the top twenty oil giants whose duties are analyzed in this part of the book. Currently, NOCs are the largest group of fossil fuel companies in terms of assets, production, revenues, and resources, as chapter 1 shows. Substantively, NOCs differ from IOCs in terms of their engagement in the provision of public services and the promotion of national social welfare, especially through social programs, infrastructure development, local procurement, and new private-sector business development. This engagement should, however, be taken with a generous pinch of salt, as most NOCs' home countries are also renowned for high levels of corruption. According to Transparency International, some NOC countries are extremely corrupt. For instance, among the 179 countries considered, the following NOC countries do not fare well (the higher the figure, the less transparent the countries are deemed to be): Mexico ranks 124, Russia 129, Nigeria 149, and Venezuela 176 in the latest 2020 Corruption Perception Index (Transparency International 2020). Possible issues of internal corruption aside, NOCs provide capital for the development of the economy and society, ensure national energy security, support national policies, redistribute social wealth, and create employment; IOCs, by and large, are focused on profits.

Despite the main differences listed above, both IOCs and NOCs face common economic, sociopolitical, and moral factors. First, the economic outlook of the oil business might be gloomy. Driven by the very rapid growth in renewable energy technologies, peak oil should occur somewhere between the second half of the 2020s and the end of the 2030s; the demand for oil has already peaked in the developed world. Peak oil is the theoretical point when fossil fuel production will hit its maximum rate, after which demand will start to drop. This notion signals a break from a past dominated by concerns about adequacy of supply from an age of (perceived) scarcity to an age of abundance and of falling oil prices that has, indeed, potentially profound implications. This issue is currently highly disputed, though. ConocoPhillips sources, for instance, argue that estimates of peak oil demand are exaggerated (Sheppard 2020); TotalEnergies in its 2020 Energy Outlook claims that oil demand will peak by 2030 and then fall to less than half of today's levels by 2050 (TotalEnergies 2020). BP CEO Bernard Looney and several analysts believe instead that peak oil demand may have occurred just before the 2020 pandemic (Raval, Nauman, and Tett 2020). The consultancy Rystad Energy (2020b) claims in its annual review of world oil resources that the pandemic will "expedite peak oil demand." According to the Organization of the Petroleum Exporting Countries' 2020 World Oil Outlook (OPEC 2020), world oil demand will instead peak in the late 2030s.

At any rate, as even IOCs, which master the production of complex asset classes at a lower cost, fear that some of their resources will get stranded, NOCs—which still post comparatively higher structural costs and show less ability in leveraging innovative field and business technologies—are at great risk of seeing their reserves become effectively worthless.

In the end, this scenario could require different kinds of regulations to curtail fossil fuel production. This can be disruptive for global oil markets as they become increasingly competitive and for major oil producing countries as they reform and adjust their economies for an age in which they can no longer rely on oil revenues for the indefinite future. In short, the oil complex and Big Oil will be deeply affected by mounting pressure targeting the supply side of climate policy.

A second group of (mostly sociopolitical) factors that threaten Big Oil is the ongoing delegitimization of the industry, given the reputational damage deriving from its increasingly acknowledged role in causing and perpetuating climate change. This seems more worrisome for IOCs, which could even see their social license to operate called into question and their financial stability shattered by the various divestment initiatives and possibly by a rapidly increasing number of climate lawsuits. NOCs, at least in the short term, do not seem to face the same risks. However, despite the latter being somewhat sheltered by protective governments and not exposed to pressure by pugnacious public opinions, it is likely that a rising global tide of social delegitimization will go some way toward shattering their security both at home and abroad.

Additionally, moral considerations—besides those specifically related to Big Oil responsibility for climate change—are becoming crucial, especially in relation to the duty of decarbonization. In fact, to favor its political feasibility, moral considerations should apply in view of protecting more vulnerable peoples, groups, and communities, including future generations, from climate impacts; to protect them from possible disruptions caused by decarbonization; and to enhance the larger transformative process needed to achieve a just low-carbon society (Patterson et al. 2018).

The achievement of the duties of reparation and decarbonization is strongly influenced by such economic, sociopolitical, and moral factors Four broad determinants that include them are outlined below. Such determinants should be considered as qualitative contextual circumstances that cannot be stipulated in absolute terms, nor are they subject to dichotomous identification (i.e., yes/no), as opposed to those introduced in the following section (assets, emissions, and responsibility) to actually group the top twenty IOCs and NOCs. The main purpose of these determinants is to qualitatively frame the quantitative categorization of IOCs and NOCs carried out in the following section and mostly to provide the backdrop for oil and gas companies to operationalize and implement their duties of reparation and decarbonization, explored in chapters 9 and 10.

The positive/negative relationships between these broad determinants and the realization of the duties are largely intuitive and in any case are justified for each single one. Furthermore, despite the trade-offs between the duties explained in chapter 6, for the sake of simplicity, no such trade-offs induced by the determinants considered are explicitly taken into account here:

1. *Societal context*: Under this label are the social, cultural, and moral issues that condemn, obstruct, and/or halt the use of fossil fuels. For instance, the emergence of anti–fossil fuel norms or even the establishment of a *fossil fuel nonproliferation treaty* (Newell and Simms 2020), the existence of a supportive coalition, the presence of counter-hegemonic forces (MacNeil and Paterson 2020), an adequate awareness of the role of Big Oil in climate change, etc. The more this societal fabric is widespread, the more likely action for establishing the duties of reparation and decarbonization becomes.

2. *Institutional strength*: Given that the duties of reparation and decarbonization in any case need political authorities, the institutional strength of the country where the company is headquartered is of great significance. The stronger the institutional context, the greater the likelihood

that Big Oil will comply with what is demanded by the duty of reparation and undertake the transition to a low-carbon world required by
decarbonization.

3. *Economic and political situation*: The richer and the less dependent
on oil revenues the homeland of the company in question is, the less
vital Big Oil's contribution to government expenditure becomes and
therefore the greater its chances are to meet its duty of reparation and
decarbonization more freely. On the contrary, the more authoritarian
the country is, the less the company has *room to maneuver*, and therefore its capacity to meet these duties is reduced. Related to this is the
scope of autonomy and the mandate that governments attribute to
NOCs, which, in fact, tends to be more limited in authoritarian countries (Victor, Hults, and Thurber 2012a).

4. *Resource availability and resource nature*: *Resource availability* basically
refers to whether the company is market-seeking or resource-seeking
since the related implications for the two duties are distinct. It is easier
for market-seeking companies to fulfill their duty of reparation since
they have a valuable asset (at least in the short to medium term, before
such resources risk getting stranded) to meet their financial obligations. At the same time, IOCs and NOCs in oil-importing countries
where consumption will keep growing are less exposed to the energy
transition, while the ability of other NOCs to adapt to the transition
might be constrained by fiscal obligations and/or social objectives
(Moody's 2020).

In order to decarbonize, resource-seeking companies should be
more likely to engage effectively in low-carbon activities, as they would
be free from the pressures experienced by those in resource-rich countries where their operations sustain the definitive owner (i.e., the host
state). On the other hand, market-seeking companies would face to a
lesser degree than resource-seeking ones the technological and infrastructural lock-in that could seriously hamper processes of decarbonization (Erickson et al. 2015a). In many respects, in the case of oil and
gas companies, particularly NOCs, the degree of dependence on fossil

fuels of their home countries can be seen as a *resource-based* lock-in When energy systems are less dependent on fossil fuels (because, for instance, they have traditionally relied also on hydropower or geothermal sources), the operationalization and implementation of the duty of decarbonization is far easier. At the same time resource availability involves technological and infrastructure lock-in—that is, as more comprehensively explained in chapter 10, the tendency of carbon-intensive systems to persist over time, despite the possibility of valid less carbon-intensive alternatives—given the long life and sunk cost of fossil fuel technologies and infrastructures (Unruh 2000).

Resource nature refers to the *geological and productive ease* of dealing with the individual fossil fuel. Typically, gas is much harder to produce and market than oil. Therefore, (mostly) oil companies such as Exxon-Mobil and Saudi Aramco would have a sizable advantage compared to (mostly) gas companies such as Gazprom. From a different perspective, the duty of decarbonization is favored by the availability in the company's host country of established sources of renewable energy (e.g., wind or year-round sunshine), since this would significantly reduce the cost of the low-carbon transition.

A useful interpretation emerging from these considerations juxtaposes to some extent IOCs and NOCs with regard to their possible achievement of the duty of reparation. IOCs have a greater likelihood of fulfilling this duty because their internal political, social, cultural, and economic circumstances are favorable to compliance with the requirements. Only point 4, *resource availability and resource nature*, indicates a different pattern: Western IOCs are resource seekers, or have fewer and by and large more difficult reserves that those of market-seeking NOCs, meaning they would be less able to rely on their fossil fuel resources to fulfill their duties. On the contrary, most NOCs' resource availability and ease should facilitate the execution of their duties.

Although a more exhaustive analysis ought to consider each company on an individual basis and in its specific context (as chapters 9 and 10 co), it seems nonetheless safe to claim that the duty of reparation in the immediate future should be a more stringent avenue of climate policy for IOCs. Only after the duty's fulfillment by investor-owned companies from the richer world should it be fully extended to NOCs and to non-Western countries. In the latter regions, the introduction of the duty of reparation should be coupled with and supported by the development of social, cultural, political, and moral issues able to forge a more favorable context as reported in point 1, societal context, above.

With regard to the duty of decarbonization, it should first be emphasized that IOCs and NOCs are clearly in different circumstances: the former do not have the social functions of the latter. Indeed, they apparently continue to respond mainly—if not only—to their shareholders' financial demands. Therefore, IOCs should aim to decarbonize to the maximum level compatible with the achievement of their duty of reparation. On the contrary, NOCs' decarbonization should take into account their role in providing indispensable revenues for their home countries; hence, their path to decarbonization needs to be more cautious to avoid undermining their social functions. The considerations carried out above therefore have more ambivalent implications for NOCs. While the societal context is expected to be unfavorable to decarbonization, the other determinants indicate different possibilities. On the one hand, quite straightforwardly, the greater the institutional strength and wealth of NOCs' home states, the more ambitious their duty of decarbonization can be. On the other hand, a market-seeking NOC may encounter high levels of resistance in a socioeconomic system largely dependent on fossil fuels and unlikely to establish supportive measures for the low-carbon transition; a resource-seeking NOC, however, may enjoy more favorable conditions in the switch to renewables.

Despite these stylized interpretations, the possibility of IOCs and NOCs achieving their duties of reparation and decarbonization in light of the determinants indicated must be examined case by case. It is worth once again remembering that the principal goal of the analysis carried out is, in fact, to provide the backdrop for examining in a contextualized way (in chapters 9 and 10) how the largest IOCs and NOCs should fulfil their duties of reparation and decarbonization.

This section rounds up the top twenty oil and gas companies in homogenous groups, according to specific objective parameters internal to the company. Coupled with the social, political, and economic factors included in the determinants outlined in the previous section, these parameters directly influence the achievement of the duties of reparation and decarbonization. The objective of the grouping process is to indicate which companies should have a similar level of requirement to duties.

A first parameter considers the assets of the company, a likely indicator of its business results given that the typical volatility of profits in the industry limits the effectiveness of their indicative capacity: the larger the assets, the greater (proportionally) their participation should be in the duties of reparation and decarbonization. In moral terms, this parameter abides with the provisions of both the ability to pay principle and the beneficiary pays principle.

The second parameter is of no small note: the company's historical contribution to greenhouse gas emissions. In principle, the greater the contribution, the larger (proportionally) its participation in the duties. On moral grounds, this second parameter would capture the requirements of the polluter pays principle.

In this context of analysis, it is useful to think of the first parameter as more significant for the duty of reparation and of the second as carrying more weight in the duty of decarbonization.

Furthermore, broader moral considerations about the responsibility of each single company are similarly important in determining to what degree each is responsible for carrying out the relevant duties. A crucial moral concern relates to the company's participation in funding, shaping, and orchestrating climate denial campaigns: if it played an active role in such campaigns (or indeed still does), its duties should be greater. This is a dichotomous entry with a simple yes or no answer: either the company has actively and for a protracted period of time participated in denial or not. The logic is that the more the company was involved in denial-related activities, the larger its participation should be in the financial rectification for the harm done and in the prevention of future harm through decarbonization. The other harm-based morally relevant facts analyzed in part I (fact A, awareness; fact B, behavior; and fact C, capacity) also concur to determine the extent of the company's responsibility. In principle, the more morally relevant fact boxes a company ticks, the larger its duties should be, as the company is testifying to greater moral responsibility for past and future climate harm. However, given the impossibility of objectivizing such facts, they are not included among the grouping parameters; rather, they will be taken into account in chapters 9 and 10 on the operationalization and implementation of the duties of reparation and decarbonization.

The twenty oil giants considered and their situation in relation to the above three parameters are listed in alphabetical order in table 7.1.

Data from table 7.1 on the twenty largest oil and gas majors can be culled to create three groups of great use in determining the individual company's morally based requirement in terms of the duties of reparation and decarbonization. The analysis then draws conclusions based on the three parameters reported to formulate the overall *ranking* of the company according to the logic outlined above. In a nutshell, the greater the assets and contribution to historical emissions and in the presence of denial, the higher the company's duties of reparation and decarbonization. It is worth recalling that this is a general indicative aggregation, which provides a useful reference for the operationalization and implementation of the top twenty oil and gas companies' duties of reparation and decarbonization discussed in chapters 9 and 10.

Having thus specified, the first group—the high-requirement (HR) one—includes BP, Chevron, CNPC/PetroChina, ExxonMobil, Gazprom, Shell, and Saudi Aramco; the medium-requirement (MR) cluster includes

Table 7.1

Parameters for mapping Big Oil

Oil and Gas Company	Assets[a]	Contribution[b]	Denial[c]
Abu Dhabi National Oil (ADNOC)–NOC	153.7	10.8	N
BP (UK)–IOC	276.5	13.8	Y
Chevron (USA)–IOC	253.8	11.8	Y
CNPC/PetroChina–NOC	608.1	14.0	N
ConocoPhillips (USA)–IOC	73.4	7.5	Y
ExxonMobil (USA)–IOC	348.7	17.8	Y
Gazprom (Russia)–NOC	352.7	35.2	N
Kuwait Petroleum–NOC	136.5	9.0	N
Lukoil (Russia)–NOC	89.6	6.7	N
National Iranian Oil–NOC	200.0	20.5	N
Nigerian National Petroleum (NNPC)–NOC	56.0	6.5	N
PDVSA–NOC	226.8	11.0	N
Pemex (Mexico)–NOC	101.8	16.8	N
Petrobras (Brazil)–NOC	229.7	6.9	N
Petronas (Malaysia)–NOC	139.5	6.2	N
Rosneft (Russia)–NOC	209.6	5.9	N
Shell (UK/Netherlands)–IOC	407.1	15.0	Y
Saudi Aramco–NOC	398.3	40.6	N
Sonatrach (Algeria)–NOC	95.2	9.0	N
TotalEnergies (France)–IOC	256.8	8.5	N[d]

[a] Oil and gas companies' assets, $ billion.

Sources: *Oil & Gas Journal* (2020a, 2020b); Annual Reports of companies; Natural Resource Governance Institute (2021), 2017 figures; ADNOC: Richard Heede, dir., Climate Accountability Institute, email communication, April 15, 2020; Lukoil: Wikipedia (n.d.a); National Iranian Oil: Wikipedia (n.d.b); Nigeria National Petroleum Corp: Amanze-Nwachuku (2007).

[b] Oil and gas companies' scope 1+3 greenhouse gas emissions 1988–2015, GtCO2e.

Source: Elaboration from the Carbon Majors Database—2017 Dataset Release (CDP 2017). According to the Greenhouse Gas Protocol of the World Resources Institute, scope 1 emissions are direct oil and gas combustions (WRI n.d.); scope 3 emissions originate from the downstream combustion (for energy and nonenergy purposes) of oil and gas distributed within the global economic system. Indeed, the largest share (roughly 90%) of oil companies' emissions consists of scope 3 emissions.

[c] Denial: Y, yes; N, no.

[d] A recent article shows that TotalEnergies knew early and concealed the truth about the relationship of its products and climate change (Bonneuil, Choquet, and Franta 2021). Its involvement in denial campaigns was however definitely more limited than that of other major IOCs; this is the reason why the company here and in the rest of the book is not considered a full-fledged denialis .

Source: Author's considerations.

ADNOC, ConocoPhillips, Kuwait Petroleum, National Iranian Oil, PDVSA, Pemex, Petrobras, Petronas, Rosneft, and TotalEnergies; and the low requirement (LR) group includes Lukoil, Nigerian National Petroleum, and Sonatrach.

A few noteworthy observations on the outcomes of this grouping process prove useful—coupled with the determinants analyzed in the previous section—for operationalizing and implementing the duties of reparation and decarbonization.

IOCs can be found in the HR and MR groups; the ones among them that, according to the analysis carried out, respond in greater terms to the morally relevant facts all belong, as expected, to the HR group. NOCs are to be found in all three groups, and major NOCs are in the HR group. Middle Eastern companies belong prevalently to the MR group, whereas African NOCs belong to the LR group. However, the largest IOCs (BP, Chevron, ExxonMobil, Shell) and NOCs (CNPC/PetroChina, Gazprom, Saudi Aramco), the *villains* in the climate change pantomime, are all in the HR group: the oil and gas companies that are generally thought to have a deeper involvement in climate change are required to be more committed to addressing it. Finally, in terms of levels of duties, there is no difference between resource and market seekers.

Indeed, the groupings illustrated in table 7.2 result from a qualitative process, which does not claim to be indisputable, nor should it be

Table 7.2
Big Oil's level of requirement to the duties of reparation and decarbonization

Level of Requirement	Oil Company
High (HR companies)	BP, Chevron, CNPC/PetroChina, ExxonMobil, Gazprom, Shell, Saudi Aramco
Medium (MR companies)	ADNOC, ConocoPhillips, Kuwait Petroleum, National Iranian Oil, PDVSA, Pemex, Petrobras, Petronas, Rosneft, TotalEnergies
Low (LR companies)	Lukoil, Nigerian National Petroleum, Sonatrach

Source: Author.

mechanically applied to other analyses of the oil industry. The outcome simply represents an attempt to map the largest companies in homogeneous groups in terms of requirements imposed by the duties of reparation and decarbonization. This process could indeed be extended to a larger pool of companies; its limitation to the top twenty suggests that this exercise is mostly exemplificative, with the objective of laying the groundwork for more thorough future investigations. A finer-grained analysis of the duties, besides considering companies' different property forms (private versus public, that is, IOCs versus NOCs), will take into account the social, political, and economic factors of the determinants highlighted in this chapter and further specified in chapter 8 as well as their different degrees of responsibility. This analysis will also weigh up other context-specific factors, such as placing more emphasis on *assets* for the duty of reparation and to *contribution* when considering decarbonization.

But before investigating these issues in chapters 9 and 10, chapter 8 provides a broader perspective on the possible approaches for destabilizing Big Oil and the oil complex. It will thus be easier to gauge and contextualize the main avenues, instruments, and mechanisms that agents of destabilization can pursue to demand that Big Oil fulfills its duties of reparation and decarbonization.

Two claims can adequately condense this chapter, the first of which was introduced among the epigraphs: "Power concedes nothing without a demand. It never did and it never will" (Douglass 1857) and "Civil society is to a great extent the only reliable motor for driving institutions to change at the pace required" (IPCC 2018, 352).

Despite the chasm of time dividing these statements and their being in reference to wholly different topics, they offer complementary insights into clarifying how Big Oil can achieve its duties of reparation and decarbonization. In fact, the first, by stating that pressure should be exerted on power in order to obtain concessions, leads to a need to identify who should exert the pressure. In this regard, the second claim—opportunely advanced by the International Panel on Climate Change (IPCC) 1.5 report—unambiguously identifies this *who* in relation to climate change: civil society, by taking the leadership, should take the helm in inducing power to change.

At the same time, the 2017 Lofoten Declaration underlines the necessity to stop fossil fuel development and to manage the decline of existing production. The document highlights how the oil industry is the nerve center of power, emphasizing the potential for a broad base of public support for disrupting the carbon-intensive status quo. The declaration affirms "that it is the urgent responsibility and moral obligation of wealthy fossil fuel producers to lead in putting an end to fossil fuel development and to manage the decline of existing production" (Lofoten Declaration 2017).

This chapter aims at analyzing how David can *defeat* Goliath. Various smaller agents with relatively limited power—by and large belonging to civil society, subnational political systems, business, research communities and other collective organizations and groups—should be able to induce a formidable agent—Big Oil—to radically change its behavior by *destabilizing* it within the oil complex as well as undermining the very foundations of the oil complex itself.

It is worth remembering that this book—in contrast to other views that postulate a spontaneous endogenous decline of the fossil fuel industry, led by farsighted industrialists (e.g., Princen and Santana 2015)—assumes that Big Oil is highly unlikely to change its behavior exclusively of its own volition. A number of exogenous forces, exerted by agents of destabilization belonging to the categories mentioned above, are expected to subvert entrenched relationships and practices of the oil complex and of the companies within this complex so as to induce change.

Given the complicated dynamics of the oil complex, in order to explore the destabilization of Big Oil, the perspective of *transition studies* (Köhler et al. 2019) must be introduced, as it can better accommodate the notions of politics and hegemony analyzed in chapter 7 into the investigation of Big Oil's destabilization. Transition studies is an area of research that scrutinizes societal systems as complex adaptive systems and analyzes them in terms of nonlinear and long-term processes of change from a transdisciplinary and integrative perspective (Avelino and Rotmans 2009). In other words, corporations and technologies are embedded within wider social and economic systems (Smith et al. 2005).

To understand how to destabilize Big Oil, it is useful to apply the basic elements of so-called transition analysis, which provides mesolevel assessments of socioeconomic agents vis-á-vis radical change in socio-technical systems (Geels, Berkhout, and Vuuren 2016). The kind of transition that concerns Big Oil can be understood as *purposive* because it involves a normative issue that tackles a problem for the common good, one that aims to achieve a set of social goals (Smith et al. 2005; Turnheim and Geels 2012). In brief, transition analysis can be employed to assess the sociopolitical acceptance and feasibility of departures from current states of affairs through investigation of the interpretations, strategies, and resources of different social forces whose alignment and coordination eventually determine the effectiveness of climate change policy.

In truth, transition analysis was originally developed—and is still exclusively used—to explore how to break free from the carbon trap and pursue a low-carbon transition (i.e., in this context, Big Oil's accomplishment of the duty of decarbonization). It can nonetheless be extended—as evinced by what follows—to include the financial rectification required by the duty of reparation among the radical changes demanded of Big Oil. At any rate, the main rationale behind the approaches included in transition studies is that in light of the previous analysis of hegemony and power, they can provide insights into the strategies and struggles of the agents of destabilization that, by triggering *sensitive intervention points* (Farmer et al. 2019), can overcome, or at least weaken, Big Oil's resistance to the erosion of its power. The main objective of this chapter is, in fact, to investigate such entry points of destabilization, the relevant agents of destabilization that are involved in them, and what steps such agents can take to put these destabilization approaches into practice.

CONFRONTING THE OIL COMPLEX AND BIG OIL'S RESISTANCE

Destabilizing Big Oil requires wearing down its resistance against attacks meant to undermine it and/or the oil complex. Resistance—largely formed through its material, discursive, and organizational hegemony—is crucial in the case of Big Oil. The anthropologist and professor of sociology David J. Hess (2014, 279), for instance, underlines that in the case of a low-carbon transition, "the political contestation by the incumbent industrial regime is so well organized that it should be at the center of the analytical framework."

Destabilization of Big Oil involves a purposive transition, as said. As a consequence, given the stressed unlikelihood that oil and gas companies have the adequate endogenous incentives to pursue social goals, external pressure exerted by agents of destabilization—mostly social movements, subnational political authorities, public opinion, and, more broadly, civil society—as well as technological advances play a pivotal role. To destabilize Big Oil, it is therefore necessary to look at social, political, and economic forces and at the sensitive intervention points through which these forces can *kick* or *shift* (Farmer et al. 2019) it within the oil complex.

Oil companies are huge economically and politically powerful and scale-intensive entities. They also have a myriad of complementary assets, such as technologies, scientific knowledge, specialized manufacturing capabilities, and lobbying skills. They have much to lose by changes imposed from the outside and enough means to buffer themselves from them. On the one hand, they are locked into their carbon-intensive business models and have sunk investments in existing technologies, skills, and people (Unruh 2000; Seto et al. 2016); they also tend to see change as risky and potentially disruptive for their existing competencies (Geels 2014).

On the other hand, as underlined in chapter 7, Big Oil has been successful in forming a stable hegemonic historical bloc—the oil complex—oriented at maintaining the current state of affairs. In particular, oil and gas companies and incumbent governments are mutually dependent with a shared interest in preserving the stability of the oil business in view of stimulating economic growth. Big Oil, in fact, depends on government to provide—or at least not upend—the general operating context (property rights, exchange rules, governance structures for corporate behavior) and for support in the form of subsidies, tariff protection, tax concession, and information and research services. By the same token, governments and socioeconomic systems (at least in all the countries of the major companies) depend heavily on economic growth and therefore systematically advance the interests of agents—in this case, the oil industry—which can further such growth as well as contribute to job creation, tax revenues, and social dynamism (Newell and Paterson 1998).

The relevance of the media to the oil complex and the latter's skill at making its voice heard have always been formidable, and yet they have experienced a boom in the last forty years. This is largely due both to the already-mentioned emergence of a neoliberal, probusiness rhetoric, which emphasizes free markets, privatization, and deregulation and managed to achieve consensual legitimacy, and to the more stringent political mobilization of corporate interests in response to social and environmental regulations. In sum, the stability of the oil complex and of Big Oil is the result of "specific alignments of material, organizational, and discursive formations which stabilize and reproduce relations of production and meaning" (Levy and Newell 2002, 87).

Therefore, to investigate how Big Oil can undergo the radical transformation in its behavior demanded by the duties of reparation and decarbonization, it is necessary to explore how to undermine the resistance to fundamental change of the core regime level alliance, that is, the oil complex.

To effectively address the climate crisis, political action and the struggle with the power of the oil industry are inevitable (Roberts 2019b). Big Oil can use different forms of power within the oil complex—which resonate with the Gramscian hegemony dimensions (organizational, discursive, and material) outlined in chapter 7—to resist changes that address the climate crisis as those required by the duties it faces.

First, organizational hegemony allows Big Oil to use *instrumental* forms of power—money, authority, access to media, lobbying skills, and networks—in direct interactions with other agents to pursue their interests and achieve their goals.

Second, such hegemony contributes as well to another form of power, *institutional*, that is embedded in political cultures, ideology, and governance structures and that greatly facilitates incumbents' resistance. For example, despite a median social cost of carbon estimated at $80–100 per tonne (Pindyck 2019), the neoliberal, promarket ideology that underpins climate governance implies that it is the market itself that should decide the best low-carbon options. Unfortunately, oil production remains so profitable that even an astonishing carbon tax of $200 per tonne would reduce global emissions by only 4 percent (Heal and Schlenker 2019). By the same token, it seems that carbon credits have provided little or no environmental gain, as they supported projects that would have come into being anyway. This is the case, for instance, for 85 percent of the projects under the United Nations (UN)'s Clean Development Mechanism, which issues a carbon credit for each ton of CO_2 avoided in the form of investments in developing countries (Cavendish 2019). Therefore, while the market-based approach to climate change may seem neutral, it actually privileges the oil industry, given its capabilities, financial resources, and established market positions. According to professor of geography Erik Swyngedouw (2010), the political dimensions that have prospered within the oil complex are camouflaged as a *postpolitical* narrative; this suggests that climate change can be addressed exclusively through techno-economic management approaches, thus excluding a wider political and cultural debate and eventually favoring the existing regime.

Third, Big Oil largely relies on *discursive* forms of power—favored by its discursive hegemony—through which it can direct and shape the narrative on fossil fuels and climate change. This is not a practice uncommon in the world of self-advantageous trade and business; however, it becomes morally problematic when it is based on false premises. As made clear in part I of the book, Big Oil not only wields formidable influence but also has become astonishingly skilled at framing the dimensions in a number of ways: diagnostic, which identifies and defines problems; prognostic, which advances solutions to problems; and motivational, which provides a rationale for action.

For instance, relying on the aforementioned forms of power, Big Oil has funded think tanks and websites engaged in climate denial whose purpose is to discourage people from fighting climate change or to drive nonexistent wedges between climate movements.

Fourth, drawing on its scientific, technical, and financial capabilities originating from its material hegemony, Big Oil has resorted to *material* power to make its technologies and activities less controversial. Most of the time, material power, in order to better prevent adverse regulations and attract potential funders, is coupled with the discursive one that proclaims that the industry's silver bullet against climate change is already in the chamber.

Examples of the use of discursive and materials powers working hand in hand are the countless technological innovations—flue gas desulfurization devices, supercritical pulverized coal technologies, coal gasification—that would have contributed to the emergence of *clean coal* and the carbon capture and storage techniques that, despite their technical feasibility and potential, still present significant uncertainties in terms of scale needed and commercial viability. Since the beginning of 2019, for instance, some of the largest oil majors (e.g., Chevron, Exxon-Mobil) have been announcing partnerships with start-ups whose technologies remove carbon from the atmosphere (Deich and Reali 2019). These examples also recall one entrenched moral question that such forms of power raise, namely the trade-off between Big Oil's potential to advance such technologies and its interest in favoring them in order to ensure an extended life to fossil fuels.

Of chief concern in relation to the destabilization of Big Oil in the context of the climate crisis are the pressures deriving from the external environment in terms of cultural and sociopolitical milieux and from markets and technologies. For instance, the emergence of anti–fossil fuel norms, climate litigation, and the reduction/elimination of subsidies to the oil industry all testify to a declining legitimacy in the first ambit; divestment and *"keep it in the ground"* initiatives as well as individual consumers' actions originate from market pressures.

In any case, politics in the broad sense, prompted by civil society activism, must erode power, and as the abovementioned IPCC 1.5 report (IPCC 2018) indicates, it is those agents that part II defines as indirect agents of justice—which as anticipated, for the dialectics of this part of the book, will henceforth simply be referred to as *agents of destabilization*—who must *play at* politics to induce Big Oil to change by eroding its powers. This is patently—albeit involuntarily—confirmed by Organization of the Petroleum Exporting Countries secretary-general General Mohammad Barkindo, who said that "unscientific" attacks by climate activists are "perhaps the greatest threat to our industry going forward" (Meredith 2019).

BIG OIL'S POWERS AND AGENTS OF DESTABILIZATION

Big Oil's power is overwhelming, but at the same time it seems to have become "promisingly unstable" (Sovacool and Brisbois 2019, 1), thanks in no small part to relevant agents' growing efforts of power erosion and subsequent destabilization.

Work on regime-destabilization evinces that by and large, destabilization processes occur in three main ways: through the progressive reduction of external financial flows; because of the erosions of legitimacy, the removal of the social license to operate, and reduced support in the external sociopolitical environment; and by means of a declining endogenous commitment of the companies themselves toward the regime they are part of (Turnheim and Geels 2013). The book acknowledges that destabilizing the oil complex is a huge challenge that involves all these modes but, in light of the analysis conducted so far, emphasizes the importance of an expressive and symbolic process of developing new social norms and institutionalizing new moral principles (Abbott 2012; Gunningham 2017) able to activate *sensitive intervention points*. Agents of destabilization are well suited to developing the transnational organized networks expected to target these entry points for eroding Big Oil's power.

In fact, agents of destabilization have different specificities and capacities and thus play different roles in undermining the various forms of Big Oil's power. This section analyzes which agents of destabilization are best suited to confront such diverse forms of powers within the politics of the oil complex. This analysis is crucial for Big Oil to achieve its (first-order) duties, since it obviates the risk of indeterminacy of agents of destabilization's second-order duties that would undermine their effectiveness, as highlighted in chapter 4.

It is first worth recalling who the relevant agents of destabilization are before investigating their role and potential in triggering the required intervention. As suggested, the most fertile ground for inducing Big Oil to change its behavior in accordance with the demands posed by its responsibility and duties is through modification of the social, political, economic, moral, and legal contexts it operates in. With regard to this point, the actions that concern the destabilization of oil and gas companies relate to external contexts (the spreading of norms and practices and the undermining of resistance to change) and to politics, the market, and technologies (financial disincentivizing and facilitation of research on and diffusion of clean technologies).

Given the importance of weakening Big Oil's resistance and the hegemonic nature of the oil complex—in which markets and technologies often seem to dance to the tune played by Big Oil, where the state provides the dance floor and the orchestra (Lindblom 2001)—it is crucial to analyze the role of these agents of destabilization that can be effective for spreading norms and undermining resistance.

Said analysis should almost take priority over other studies into the actual destabilization of Big Oil; understanding the role of such agents may represent a sort of unavoidable prerequisite to exploring the role of the agents involved in more operational tasks such as financial disincentivizing and the search for new technologies. The latter depend largely on government (e.g., fiscal, regulatory, industrial, legal instruments as well as funding for research) and market initiatives that are more conceivable when the sociocultural context is *ripe*, as pointed out in part II. However, to achieve this level of maturity, the main actions must be rooted in norm spreading and resistance undermining.

For the sake of clarity, agents of destabilization involved in spreading norms and undermining resistance are considered to be *primary* forces; those who use regulations, markets, legal action, and/or financial means to steer/change Big Oil's behavior are referred to as *operational* forces. Although the distinction between primary and operational agents of destabilization is blurred, since they both can focus on the same sensitive intervention points, primary agents tend to aim at shifting the dynamics of a system by substantially changing its rules and trajectories (e.g., key values and concepts and institutions in the socioeconomic-political context). Operational agents, on the other hand, are those who kick a system into novel trajectories, based on the changed underlying system dynamics or without introducing new ones (Farmer et al. 2019).

Primary agents of destabilization lay the fundamental groundwork in a bottom-up and quasi-anarchic way to prepare societies to acknowledge and accept the inadmissibility and future impossibility of fossil fuel. To adopt a parallel trajectory to society's gradual rejection of tobacco, these agents should also raise awareness of the negative health effects, both locally and globally, of fossil fuel combustion. On this fertile ground, operational agents of destabilization should introduce the measures aimed at challenging Big Oil's powers in view of its destabilization.

This section develops a general framework of the agents of destabilization who can most expect to erode Big Oil's powers by exerting fundamental pressure on the oil complex and also briefly highlights their roles with regard to the climate crisis. The following two sections analyze the nature, objectives in terms of sensitive intervention points, and strategies of primary and operational agents for Big Oil destabilization.

In light of the multifaceted features of Big Oil's powers and with reference to the indirect agents—those indicated as having second-order duties to ensure that reparation and decarbonization are enacted—it seems that charismatic individuals and social movements are in the most advantageous position to fracture the instrumental, discursive, and institutional forms of Big Oil's power. Research institutions and financial actors are the most effective forces in challenging its material power.

As for the first forms of power (instrumental, discursive, institutional), charismatic individuals have a prominent role in calling on societies and their broad cultural contexts to respond to the challenges posed by the climate crisis and in helping develop the adequate institutional responses. It could be the novelty of the challenge that makes people particularly susceptible to compelling outsiders: given the extremely limited permeability of the hegemonic bloc to external interference, these charismatic individuals must not hail from within the established hierarchy of the group so as to avoid any finger-pointing for responsibility for the status quo.

Such front-runners in the climate discourse are, by and large, able to quickly mobilize people to consider and confront particularly urgent aspects of the climate crisis, such as the harmfulness of fossil fuels (e.g., Pope Francis, Greta Thunberg), the reckless behavior of Big Oil, and the urgency to divest from it (e.g., the environmental activist Bill McKibben). These agents prepare the ground to monitor the industry's efforts to fulfill its duties of reparation and decarbonization. Their distinctive role is to converge and catalyze the pressures that hail from civil society, transforming them into an ever-increasing wave of novel forces to challenge Big Oil's power especially on instrumental, discursive, and institutional grounds.

These novel forces are usually referred to as *social movements*, that is, "networks of informal interactions between a plurality of individuals, groups and or organizations, engaged in political or cultural conflicts, on the basis of shared collective identities" (Diani 1992, 1). For practical purposes, social movements are coalitions of individuals and organizations from both civil society and the private and public sectors. Social movements' collective action is indeed a key factor to changes in human, social, and economic systems (Della Porta and Diani 2006, 33–63); they are the foremost primary agents of destabilization for weakening Big Oil's resistance: "Challenging and undermining the fossil fuel historical bloc on the scale necessary for maintaining the familiar stability of the Earth system will surely rely on the success of widespread and sustained movement building" (Phelan, Henderson-Sellers, and Taplin 2013, 216). For instance, BP, in a leaked briefing dated January 14, 2020, titled *BP Creative Workshop*, clarifies that a major threat to the company comes from climate movements and, in particular, their capacity to erode the social license to operate of oil companies (BP 2020a).

Collective action, however, is not automatically triggered by the structural tensions generated by and within the oil complex. Rather, a number of factors determine whether or not collective action occurs. Among the most prominent are the availability of adequate organizational resources, the ability of movements to create appropriate ideological and practical representations of the issues at stake, and the presence of a favorable context. In relation to the oil complex, such factors face a particularly harsh obstacle: the stubborn opposition and tenacious resistance of the Big Oil historical bloc.

As said in chapter 7, most of the social movements working on climate change established themselves as counterhegemonic forces through the representation of climate change as a threat to humanity. Its main protagonists—including oil and gas companies—are presented as being mostly concerned with safeguarding their own interests, often in conflict with those of humanity. By so doing, social movements have successfully begun to threaten the contingent hegemonic stability of Big Oil's dominant position.

The counterhegemonic forces of social movements are eroding Big Oil's instrumental, discursive, and institutional powers along, at least, three main avenues: first, by exerting social, political, and economic pressure to endorse the institutional divestment of assets including stocks, bonds, and other financial instruments connected to these companies; second, by demanding an immediate stop to new fossil fuel development—in terms of both production and infrastructures—and that the oil industry move toward a just low-carbon transition; and third, by spreading awareness—despite the mighty rhetorical denial machine and action of some of the oil majors, as depicted in part I—that climate change is occurring in the here and now and that it is going to wreak severe consequences globally if it is not addressed promptly.

The disruption of Big Oil's material power will require the efforts of research institutions to play the all-important role of being the genesis of technological and social innovation. Their ability to weaken Big Oil's material power mainly lies in developing new products, services, and business models; contributing to creating markets for novel technologies; and diffusing such technologies. At a different level, research institutions should also contribute to shaping societal discourses and problem framing, lobby for specific policies and regulations, develop industry standards, legitimate new technologies, or shape collective expectations (Binz et al. 2016; Rosenbloom, Berton, and Meadowcroft 2016).

On the other hand, financial actors—mostly pension funds and sovereign wealth funds as well as central, investment, and commercial banks—should modify their objectives and practices, especially in view of the fact that their cash injections into the industry have been partly to blame for the recent growth spurt of oil and gas companies (RAN et al. 2021). In order to erode Big Oil's material power, financial institutions should first and foremost take the very simple step of abolishing all fossil fuel–related funding. In broader terms, they should also adopt a number of internal measures, such as strengthening the assessment and monitoring of climate-related financial risks, integrating sustainability into their own portfolio management, and sharing knowledge with other stakeholders on the management of climate-related financial risks (Carney, Galhau, and Elderson 2019).

Indeed, the relationship between Big Oil's powers and agents of destabilization is very useful in analytical terms, but by no means is it clear-cut. For instance, specific agents of destabilization might have a distinctive role with regard to particular forms of power: financial agents play a decisive role in weakening Big Oil's institutional power, despite the fact that a common theme between this kind of power and the discursive and material ones is the centrality of charismatic individual and social movements. Or, again, some forms of power are mutually fortifying, and therefore their erosion requires the concerted pressure of multiple agents of destabilization: for example, material and discursive powers often strengthen one another, and thus both social movements and financial agents are decisive in addressing these forms of power. A further critical issue relates to the impossibility—or extreme difficulty—of agents of destabilization, both primary and operational, having an active role in authoritarian regimes: unfortunately, the great majority of NOCs belong to countries under authoritarian regimes (Economist Intelligence Unit 2021).

PRIMARY AGENTS OF DESTABILIZATION

The current historical juncture has brought climate change to the fore-front of global news, meaning that the time seems ripe to finally address it: while its impacts are increasingly evident worldwide, climate concern has also trickled up to the top of the political agenda. However, in other circumstances during the past three decades, it also seemed that our political leaders were on the cusp of seriously engaging with climate change. Alas, for a variety of reasons, they did not tackle the issue. One fundamental differentiation that can be made with the past is that previous responses to the climate challenge were fundamentally built on the evidence of the mounting environmental crisis, while nowadays, despite the enduring apparent helplessness of UN multilateral negotiations, action gets momentum by the mobilization of people, particularly to destabilize the oil complex and Big Oil.

To engage the public and give them the prerequisites to challenge entrenched powers, it is essential to change human values and cultural worldviews about climate change, which are the *currency* of persuasion (Sovacool and Griffiths 2020). In the context of such a politicized issue in the so-called posttruth world, where the political culture is framed more on appeals to emotions than on facts, the latter have become largely irrelevant to the public. Modifying values and worldviews about climate change and creating the conditions for people and other agents to act and be listened to by decision makers is, in a nutshell, the role of primary agents of destabilization.

The great achievers in this daunting task, as said, are the charismatic individuals and social movements that are acting either as norm entrepreneurs or champions. After the failure of the 2009 COP 15 in Copenhagen, charismatic members of civil society stepped into the guise of what can be effectively described as norm entrepreneurs. For example, the prominent campaigner Bill McKibben has mobilized moral outrage against fossil fuel companies. Similarly, Pope Francis (2015) called for the phasing out of fossil fuels in his climate change encyclical; more recently he asked "the world to give up fossil fuels," claiming that the climate is in a state of "emergency . . . caused by human activity" (Cummings McLean 2019); encouraged governments and corporations around the world to urgently address climate change (Knutson 2020); and eventually exhorted Catholics to divest from fossil fuels (Pullella 2020).

Other charismatic individuals have emerged from civil society. The international lawyer Polly Higgins, for example, has long striven to have ecocide recognized as an international crime. This, according to Jojo Mehta, cofounder of the Stop Ecocide campaign, would help to create a cultural shift in how the world perceives acts of harm toward nature (Mehta and Jackson 2021). This crime would make the people who commissioned it—such as oil majors' CEOs and executives and, within a state context, ministers and heads of state—liable for the harm they do to others. Higgins was particularly targeting Shell to establish whether the company could be accused of ecocide (Hope 2019b).

The teenage Swedish activist Greta Thunberg, who has successfully managed to mobilize millions of mainly young people to take part in *Fridays for Future* protests across the globe, warns that climate change is generating an existential crisis for humanity, in particular for future generations, and that too little is being done, urging decision makers to listen to scientists (Sabherwal et al. 2021).

Social/moral norms are for the most part originated by *norm entrepreneurs*, agents—often individuals such as the ones mentioned above— highly motivated to overcome a perceived injustice/problem/barrier, usually through an organizational platform, such as nongovernmental organizations, social movements, and international organizations (Finnemore and Sikkink 1998). Norm entrepreneurs aim to forge a new standard of behavior, one that will be accepted in the international system. Given the vested interests that typically coalesce around the injustice/problem/barrier that norm entrepreneurs wish to tackle, they must adopt creative tactics to disrupt existing logics. *Norm champions* are those political and nonpolitical agents that promptly adopt a norm and, through national and international channels, pressure others to do the same. Norm entrepreneurs and champions are usually linked through transnational advocacy networks that campaign for change at multiple levels (Keck and Sikkink 2014).

By and large, the challenge to Big Oil's power greatly benefits from charismatic norm entrepreneurs that include figures such as religious leaders, film actors, writers, and other gifted and dedicated communicators. These charismatic individual agents of destabilization are often the best channels to reliably and effectively communicate climate change to the public. Climate scientists able to adopt accessible language can play a significant role too, such as the German, Austrian, and Swiss Scientists for Future initiative, as can science journalists writing for mass appeal publications. In broader terms, an interesting insight into individual primary agents of destabilization is provided by the Climate 100 list (the world's most influential people in climate policy) published by Apolitical, a global network for public servants and government, that includes household names from the spheres of broadcasting, politics, journalism, and academia (Apolitical n.d.). By the same token, environmental advocacy groups and reliable investigative media sources are also important primary agents that can step into the role of both norm entrepreneurs and norm champions.

Among the nongovernmental organizations most active in their advocacy of the responsibility of the oil industry and its duties to rectify the harm caused by climate change are the Union of Concerned Scientists (UCS), the Center for International Environmental Law (CIEL), and Oil Change International. For example, in 2017, CIEL published a fundamental report—*Smoke and Fumes: The Legal and Evidentiary Basis for Holding Big Oil Accountable for the Climate Crisis*—that evaluates the evidence for Big Oil's liability in light of fundamental principles of legal responsibility, concluding that oil and gas companies should be held accountable for climate harm (CIEL 2017).

At the same time, respected newspapers and magazines (e.g., *The Guardian*, *The New Yorker*) often publish pieces both by their journalists (e.g., George Monbiot and Elizabeth Kolbert) and external experts (e.g., Peter Frumhoff of UCS, Bill McKibben, the historian of science Naomi Oreskes) supporting the thesis of responsibility and financial liability for Big Oil. Similarly, investigative journalism (e.g., the *Climate Investigation Center*, *DeSmog*, and *Inside Climate News*), while casting light on the concealed aspects of the oil world, is repeatedly making the case for financial rectification. And in the case of *The Guardian*, advertising from the fossil fuel industry has been banned.

An example of an apparently unlikely yet symbolically very significant norm entrepreneur with regard to divestment from fossil fuel is the leading peer-reviewed medical journal *BMJ* (formerly the *British Medical Journal*). It launched a divestment campaign—dubbed *investing in humanity* based on a case for divestment made previously by the same journal (Tillmann et al. 2015)—aimed at health professionals and medical organizations using moral arguments to justify its campaign (Abbasi and Goodlee 2020). In the same editorial, the *BMJ* announced also that it will no longer accept advertisements from fossil fuel companies or publish research funded by them.

An important global social movement that acted as a norm champion is the Fossil Free divestment movement, which is at the forefront of civil society initiatives to raise public consciousness about the need to decarbonize socioeconomic systems by divesting from the business. As already underlined, according to the movement's website, as of September 2021, 1,335 institutions are divesting $14.65 trillion from the fossil fuel industry, while more than 58,000 individuals are divesting $5.2 billion (Fossil Free n.d.).

A further relevant norm champion social movement is one repeatedly mentioned in these pages, the Keep It In The Ground movement, about which, interestingly, the *Wall Street Journal* claims, "What was the radical-left position of a few years ago—*Keep It In The Ground*—is now mainstream" (Strassel 2019). Other important social movements are the Powering Past

Coal Alliance, which includes more than twenty-five countries that have pledged to phase out coal-fired power generation; 350.org, whose objective is to end the age of fossil fuels and build a future free from the destructive impacts of climate change and from the out-of-control corporations that caused it; and the Stop the Money Pipeline coalition, which demands that banks, asset managers, insurance companies, and institutional investors stop funding, insuring, and investing in fossil fuel infrastructure.

A particularly interesting—yet powerful despite its short history—international movement is the so-called Fridays for Future mentioned above. It was initiated in August 2018 outside the Swedish parliament when Greta Thunberg held a sign that read *Skolstrejk för klimatet* (School strike for the climate). This movement was initially made up of a handful of school-age students who did not attend classes on successive Fridays to take part in demonstrations to demand action to prevent further climate change. In a very short span of time, the movement brought to the streets millions of protestors worldwide of all ages and from all walks of life. Thunberg is the charismatic norm entrepreneur who targets adults in positions of authority, both in fossil fuel corporations and political institutions, for their responsibility for carbon emissions and doing far too little to reduce them. In the same vein, for instance, is the American Sunrise Movement, a youth-led political movement that advocates political action on climate change.

The initiatives promoted by the norm champion Extinction Rebellion (XR), a sociopolitical movement established in the United Kingdom in May 2018, seem to be making an impact. At the time this book was being finished, XR was operational in eighty-three countries and had 1,196 local groups. XR uses civil disobedience and nonviolent resistance to protest against the climate crisis and defines itself as "a decentralised, international, and politically non-partisan movement using non-violent direct action and civil disobedience to persuade governments to act justly on the Climate and Ecological Emergency" (XR n.d.).

Casting an eye in particular on the duty of decarbonization, anti–fossil fuel norms are expected to form the necessary cultural and moral backbone of awareness about the harm wreaked by fossil fuel–related activities, with the objective of favoring socioeconomic systems' transition toward less carbon-intensive models. Therefore, such norms aim at changing current socioeconomic systems by convincing people intellectually and emotionally that fossil fuels, given their harmfulness, are morally wrong. In the case of the duty of decarbonization, the moral-normative contents of such norms should consist of the prohibition of operating with fossil fuels.

In brief, with regard to decarbonization, while individual norm entrepreneurs remain crucial, the absence of adequate political action at the national and international levels currently makes social movements the most powerful anti–fossil fuel norm champions for effectively overcoming the carbon era. Their mobilization is largely focused on the approaches of supply-side climate policy described in chapter 6, with the goal of restricting fossil fuel supply so as to steer societies toward a carbon-free world. At any rate, social movements and, more broadly, nonstate actors have had a remarkable impact on reducing carbon emissions worldwide (Hsu et al. 2019).

Primary agents of destabilization have important normative and practical implications for Big Oil and climate policy and politics. On normative grounds, they shed light on different climate policies focused on the supply side, much more relevant to engaging Big Oil, as said, in the climate change struggle. On practical grounds, the broadness and inclusiveness of such agents (especially social movements) make the fight against climate change a truly global effort at any societal level.

In sum, on the one hand, social movements—most often started, organized, and led by charismatic individual norm entrepreneurs—capable of laying the groundwork to destabilize Big Oil have a broader goal of societal transition by stopping unwanted practices and policies, leveling out social inequalities, and promoting low/zero-carbon technologies while supporting alternative just measures. On the other hand, social movements have been successful in integrating the previously dispersed individuals and coalitions, thereby increasing their efficiency. All in all,

primary agents of destabilization have thrown open the doors of the pol
icy process by encouraging greater participation from different coalitions
in decision making and in policy implementation. By so doing, they
have managed to exert substantial destabilizing pressure on oil and gas
companies—social movements can actually significantly orient votes in
elections and mobilize extensive street protest—and prepare the terrain
for the implementation of incisive actions required to compel Big Oil to
meet its duties of reparation and decarbonization.

OPERATIONAL AGENTS OF DESTABILIZATION

The quintessential operational agents of destabilization in the oil complex
are political authorities at various level, which—thanks to the *fertilization*
seed work done by primary agents of destabilization—are expected to
effect changes by introducing the necessary actions to disrupt Big Oil. For
instance, political authorities can implement instruments—both regulatory and market-based—to limit the supply and demand of fossil fuels,
eliminate subsidies to the oil industry, and ban fossil fuels altogether: in
this spirit, US president Joe Biden signed an executive order in January
2021 directing federal agencies to eliminate subsidies for fossil fuels.

While a thorough analysis of the role of political authorities in climate
change would be impossible, with regard to the oil industry it is useful to
shed light on the role of political agents at the subnational level. The following two chapters, however, will include some national-level actions in
relation to the duties of reparation and decarbonization for the different
groups of oil and gas companies. Subnational political agents have cham-
pioned anti–fossil fuel norms by, for instance, banning fracking in their
jurisdiction or, more ambitiously, phasing out fossil fuels, as happened
in Hawaii and stated in the bill signed by California's former governor
Jerry Brown, which should be followed by other US states, from Nevada
to Michigan to New York, as well as Washington, DC (Roth 2019).

In the United States, subnational political authorities have acted as
operational agents of destabilization in a broader and possibly more

effective way by suing Big Oil through state lawsuits alleging both consumer and investor fraud over climate risks (Drugmand 2019a, 2019b), a
strategy that calls to mind the public proceedings that forty-six US state
attorneys general collectively launched against the tobacco industry in
1999. For instance, the District of Columbia and Hawai'i's Maui County
sued oil majors BP, Chevron, ExxonMobil, and Shell, accusing them
of deceiving consumers about climate change risks and engaging in a
coordinated decades-long campaign to mislead the public (Savage 2020).
The State of Minnesota instead filed a lawsuit against Exxon, three Koch
Industries entities, and the American Petroleum Institute, accusing
them of consumer fraud and other violations for their protracted deception initiatives (Hasemyer 2020).

It is interesting to note how the establishment and instigation of such
legal initiatives are greatly favored by primary agents of destabilization.
For instance, New York City (a subnational political authority) mayor Bill
de Blasio (a charismatic individual), presented the city's lawsuit seeking
billions in damages from five major oil companies (BP, Chevron, ConocoPhillips, ExxonMobil, and Shell) to cover infrastructure improvements
needed to protect New Yorkers from the increasing effects of climate
change. During the press briefing, he clarified that the industry must be
held accountable and financially liable for harm caused by climate change.

Climate litigation is rapidly growing worldwide: more than fifteen
hundred lawsuits, many targeting governments or corporations, have
been filed in thirty-seven countries; cases are increasing in number outside the United States, including the Global South, which has seen fifty-
eight cases so far (Setzer and Higham 2021).[1]

Social movements are actively exploring innovative approaches to hold
Big Oil accountable on legal grounds. Exploiting various legal loopholes
in the United States, a vast number of shareholders are currently plaintiffs against oil and gas companies, suing companies' officials, directors,
and board members for not having protected their investments and the
company from climate risk (Savage 2019). The industry is, in fact, starting to acknowledge that climate litigation threatens its business; Shell,

for instance, in its 2018 *Annual Report* writes: "Further, in some coun tries, governments, regulators, organizations and individuals have filec lawsuits seeking to hold fossil fuel companies liable for costs associatec with climate change. While we believe these lawsuits to be without merit losing any of these lawsuits could have a material adverse effect on oui earnings, cash flows and financial condition" (Shell 2020a).

Not all lawsuits are successful: as already said in chapter 5, the New York State attorney general's case was dismissed by the judge, ruling that the company did not defraud investors out of up to $1.6 billion by covering up the true cost of climate change regulation. At any rate, climate litigation is becoming increasingly frequent in all four corners of the globe, and generally the main objective is to make fossil fuel companies liable for the impacts of climate change. In fact, courts have often helped to accelerate social change in critical moments in history—the end of slavery, racial desegregation, gender equality—and it is no surprise that they are in demand to contribute to solving the climate crisis.

Other operational agents of destabilization who have been promisingly active in confronting the oil complex are the economic ones: the title of a piece published in the *New Yorker* by McKibben (2019) is as direct as it is compelling: "*Money Is the Oxygen on Which the Fire of Global Warming Burns.*" The scientific and policy community has even provided principles to cut off the oxygen the industry thrives on—the three *Oxford Martin Principles for Climate-Conscious Investment*, commitment to net-zero emissions, profitable net-zero business model, and quantitative medium-term targets—for assessing whether investments are consistent with long-term climate goals (Millar et al. 2018).

Global finance could be the driving force behind the new phase in the effort to address climate change, as investors increasingly become aware of the reality of climate risk; in an attempt to protect their interests, they are channeling investments into greener ventures, consequently stimulating climate stability (Colgan, Green, and Hale 2020). The European Central Bank and the US Federal Reserve, for example, have signaled their intent to make climate considerations a central part of finance. In

November 2020 Mark Carney, the former governor of the Bank of England and current UN special envoy for climate action and finance, called on banks, insurers, and investment funds to disclose how closely their business choices were aligned with climate goals as part of the economy-wide transition to the hoped-for net-zero aims (Kirka 2020).

Others warn instead that the role of financial institutions should not be overstated because sometimes they offer *various greenery* to their clients to justify charging higher fees (Economist 2020). On a more sour note, it is worth noting that investments in initiatives to reduce carbon emissions fell in 2018 and that a percentage of it saw its way back to the fossil fuel industry (Buchner et al. 2019).

Economic agents such as commercial banks, development banks, insurers, pension funds, and sovereign wealth funds are the main drivers behind growth in committed assets in fossil fuels. They must facilitate the development of capital market instruments that package risk and return and asset allocation strategies that align portfolios with the low-carbon transition. In this vein the 2019 UN Trade and Development Commission report demanded a profound restructuring of the global financial system to cope with climate change (UNCTAD 2019).

Commercial banks are already starting to get into line. For instance, JPMorgan Chase, the world's biggest fossil fuel investor, committed to a Paris Agreement alignment of its lending practices (Benoit 2020) and announced that between 2021 and 2030 it would finance and facilitate more than $2.5 trillion to advance climate action and sustainable development. Similarly, Deutsche Bank, Citigroup, and Barclays (which have more than $47 trillion in assets) are among the 130 banks that adopted the new UN-backed responsible banking principles to fight climate change that requires their loans be shifted away from fossil fuels (Green 2019). UniCredit, the biggest Italian bank, has pledged to halt all lending for coal projects by 2023 (Za 2019). Even the US giant in investment banking, Goldman Sachs, announced that it will no longer finance oil drilling or exploration in the Arctic and coal mining and coal power projects worldwide (Brown 2019).

Unfortunately, the reality still looks grim: the world's biggest sixty banks have provided $3.8 trillion of financing for fossil fuel companies since the Paris Agreement, and the slump in energy demand caused by the COVID-19 pandemic did not stop this upward trend, seeing as the figure for 2020 was higher than those of previous years (RAN et al. 2021). In fact, most of directors at the world's biggest banks have affiliations with the fossil fuel industry (Cooke at al. 2021).

Development banks, such as the African Development Bank, the Asian Development Bank, the Asia Infrastructure Investment Bank, the European Bank for Reconstruction and Development, and the World Bank at the global level are similarly planning to divest from fossil fuels (Jerving 2019). In the meantime, the European Investment Bank, which defines itself as the "lending arm of the European Union" and "the biggest multilateral financial institution in the world and one of the largest providers of climate finance," announced that it will stop financing fossil fuel energy projects from the end of 2021; that its future financing activities will focus on the promotion of clean energy innovation, energy efficiency, and renewables; and that it will mobilize €1 ($1.13) trillion into climate action and sustainable investment in the decade to 2030 (European Investment Bank 2019). The European Investment Bank adopted the Climate Bank Roadmap through which to increase its lending to climate action and green activities to more than half of its funding activities by 2025 (Farand 2020). All the while, the European Central Bank has announced that it will phase out climate-warming investments in favor of green bonds (Farand 2019).

By the same token, institutional investors are playing an increasingly important role. For instance, the investor initiative Climate Action 100+, which numbers more than 360 investors with more than $34 trillion worth of assets under management, aims to engage with the major carbon-emitting companies they hold shares in so as to address the climate risk. A Climate Action 100+ shareholder resolution to get BP to demonstrate that its strategy was consistent with the goals of the Paris Agreement was approved by the British oil giant's board and is now legally binding (Espiner 2019). Among Climate Action 100+'s influential institutional investors is the Church of England's property asset body, the Church Commissioners; it is worth noting that faith institutions and religious groups form the largest bloc within the global divestment movement (Dodd 2019). By the same token, the insurance sector continues to lead the trend of divestment, with over $3 trillion in assets committed. In addition, insurance companies are keeping a close eye on how climate change harm is affecting their business interests and appear to be readying themselves to file lawsuits against the fossil fuel industry, seeking to recuperate payouts to policyholders for climate damages (Sullivan 2019).

The American stock guru Jim Cramer affirmed that "I'm done with fossil fuels. They're done" (Pound 2020). Basically, he claimed that fossil fuel stocks have become washouts because the divestment movement is forcing people to dump them. Indeed, after years of hesitation and unheard shareholder resolutions, some funds are trickling out of the oil markets too. Sovereign wealth funds and pension funds are similarly abandoning fossil fuels: Ireland's €8.9 ($10.1) billion sovereign development fund is committed to divesting from fossil fuels, Norway's $1 trillion wealth fund divested from 150 companies active in the exploration and production of oil and gas (Davies 2019), and Denmark's MP Pension Fund divested from 24 oil majors an overall value of $133.9 million (Baker 2020). Similar initiatives are being undertaken by over 100 globally significant financial institutions that have divested from coal, including 40 percent of the top 40 global banks and 20 globally significant insurers (Buckley 2019). One of the world's leading asset managers has even warned that Big Oil's directors must act on climate change; otherwise, they risk being voted out (Greenfield 2019). The asset manager Legal & General Investment Management unsuccessfully tried to convince ExxonMobil—as one of its top twenty shareholders—to better address its climate risk. As a result, in June 2019 Legal & General Investment Management announced that it had divested approximately $300 million worth of its Exxon shares and would use its remaining stake to vote against the reappointment of ExxonMobil chairman and CEO Darren Woods (Giblon 2019). Black-Rock, the world's biggest asset management firm, has lost an estimated $90 billion over the last decade by ignoring the serious financial risk of investing in fossil fuel companies. Multibillion-dollar investments in the world's largest oil companies—including BP, Chevron, ExxonMobil, and Shell—were responsible for the bulk these losses (Ambrose 2019). Additionally, in early 2020 BlackRock announced that it would stop investing in thermal coal (Rowell 2020) and later that it would sell shares in the worst climate polluters, even if it still manages $85 billion of coal assets and its investment in coal producers with expansion plans exceeds $24 billion (Cuvelier and Pinson 2021). In the meantime, investment funds that divested from fossil fuels profited not only morally but also in terms of financial return (Sanzillo 2021).

Divestment is an effective strategy for destabilizing Big Oil through the erosion of its material power. It was inspired by the perceived success of the 1980s South Africa divestment campaign to pressure the South African government into ending apartheid. Yet due to the multiple secondary effects and the market and political uncertainties, the ultimate effectiveness of divestment in reducing emissions is up for debate, so the goals of fossil fuels divestment campaigns need to be specified. Contrary to commonsense intuition, divesting merely shifts the ownership of a (publicly traded) company without actually altering the flow of funds in or out of it; as a consequence, in the short term the underlying economics of a company is largely unchanged by even a loudly proclaimed divestment. The company does not suffer major financial loss, and its decision making should in principle remain unchanged. In brief, the market value of the company is irresponsive to divestment, making its short-term effectiveness in *punishing* Big Oil or reducing emissions very limited. The direct impact of divestment—albeit a reduced one—on the valuation of fossil fuel companies can instead be found in changes in market behavior or in constrained debt markets. In the first case, divestment may close off channels of previously available money, thus initiating downward pressure on the stock price of a targeted firm. Second, in poorly functioning markets and in countries with low financial depth, divestment can diminish the pool of debt finance for Big Oil and increase discount rates.

It is, however, more useful to understand divestment as a long-term strategy based on the stigmatization of the industry with three main objectives: to force companies to stop the use of fossil fuels, to pressure them to undergo *structural change* that will lead to a drastic reduction in carbon emissions, and to urge governments to pass legislation, such as bans on further drilling or a carbon tax. The divestment stigma generates several negative consequences for oil and gas companies. It can result in customers, suppliers, and potential highly skilled and qualified employees running scared and can induce governments and politicians to engage only with *clean* companies to prevent adverse effects damaging their reputation or endangering their reelection. Shareholders can demand changes in management of companies; stigmatized companies can also be excluded from public tenders, acquiring licenses or property rights for business expansion, or be weakened in negotiations with suppliers and be denied new contracts or mergers/acquisitions. Of far greater significance is the fact that stigmatization can impact Big Oil through new legislation: nearly all non–fossil fuel-related divestment campaigns managed to successfully lobby for restrictive legislation affecting stigmatized firms.

All these factors greatly increase the uncertainty of future cash flows for the stigmatized company, thus compromising its market value and eventually even its operativity: it is this aspect that forms its long-term *punishment*. In synthesis, while divestment campaigns do not seem to have a significant direct impact on reducing emissions, divestment as an institutional strategy can help drum up the necessary support for a climate agreement and effective climate policies in the medium and long term (Ansar, Caldecott, and Tilbury 2013; Braungardt, Bergh, and Dunlop 2019). Divestment causes panic among oil executives, who pool their public relations resources to combat the increasing number of steadfast divestment movements emerging worldwide (Farand 2018). Indeed, the previously mentioned guru Jim Cramer, arguing in favor of fossil fuel divestment, compared fossil fuel stocks to the stigma attached to investing in tobacco companies, saying they are in the "death knell phase" and adding, "They're tobacco. I think they're tobacco" (Pound 2020).

A final consideration should go to agents hailing from the worlds of both technology and R&D—developing low-carbon technologies and, more broadly, working at the innovation frontier of fuels and energy production—who play a relevant part in destabilizing Big Oil. Without overemphasizing their role—unlike *old-fashioned* innovation studies, which placed great faith in the impact of innovations—they can deliver disruptive low-carbon technologies at a comparable or, increasingly, in some cases, inferior cost to that of consolidated fossil fuel technologies. Alternatively, they can disrupt the fossil fuel world from the inside. For instance, about one thousand Australian engineers and ninety organizations are pressing engineering firms to abandon fossil fuel projects, especially the most controversial coal-related ones (Smee 2019).

EXOGENOUS SHOCKS AND BIG OIL'S DESTABILIZATION: CLIMATE CHANGE AND FINANCIAL CRISES

According to the 2021 Doomsday Clock Statement of the *Bulletin of the Atomic Scientists,* "the existential threats of nuclear weapons and climate change have intensified in recent years" (Mecklin 2021). Firsthand experiences of the lethal effect of nuclear weapons—from their first tests at the Trinity Site in New Mexico in the United States on July 16, 1945, through to the harrowing massacres of Hiroshima and Nagasaki in early August of the same year to the countless subsequent tests in the atmosphere, underwater, underground, and in outer space—instilled such terror into humanity that between 1965 and 1968, leaders were forced to sign a nuclear nonproliferation treaty that prevented an escalation in the development of nuclear weapons and weapons technology, even at the height of the Cold War standoff.

Continuing with this parallelism, unfortunately, firsthand experience of the impacts of the climate crisis and wider acknowledgment of its dire evolution have not so far managed to provoke a similar emotional reaction in humanity or induce decision makers to seriously address climate change. Possibly, increasingly frequent and more extreme weather events may prompt an adequate response to destabilize Big Oil: a climate emergency mobilization of people, technology, and policy is the most likely combination to enact the change required (Gilding 2019). Scientific studies on this eventuality remain, however, inconclusive. For instance, one work (Dixon, Bullock, and Adams 2019) found that emphasizing the role of climate change in natural hazards (hurricanes, wildfires, and blizzards) that resulted in significant loss of life and property produces unintended effects on (American) people, who build up a kind of resistance to the news and a reduction in the perceived severity of the hazard, often dubbed *"compassion fatigue"* and *"apocalypse fatigue"* whereby the limits of emotional attention and empathy are stretched too thin by overexposure to shocking news, events, or calls for support and where a *"seen it all before"* mentality starts to ebb away at initial dismay. Another study (Boudet et al. 2020) evinces that although any single event may have limited impact on discussion or collective action about climate change, what is of vital importance for mobilizing people and spurring action is partisanship and the attribution of the event to climate change. A further study (Bergquist, Nilsson, and Schultz 2019) shows that experiencing climate-related disasters firsthand (the study specifically concerns the experience of the hurricane Irma by residents of Florida) magnifies negative emotions toward climate change, strengthens people's beliefs that the disaster was actually caused by climate change, and encourages a willingness to make personal sacrifices in order to protect the environment.

More studies are required to be able to uncontroversially presume that the current and prospected impacts of climate change can induce civil society to exogenously shake up the oil complex and destabilize Big Oil—unless, unfortunately, a major global climate-related disaster tips the balance in favor of a radical change in the world order.

From a different perspective, climate change could induce the world economy to spiral into a global financial crisis similar to the one triggered in December 2007. Such a crash would be an exogenous factor that could significantly disrupt the oil complex and destabilize Big Oil.

Indeed, nowadays most people accept in principle that climate change threatens financial stability, as a report commissioned by the US Commodity Futures Trading Commission warns (Davenport and Smialek 2020), thus incurring substantial public costs, and that financial regulators (e.g., central banks and governmental financial stability mechanisms) should do their part in safeguarding the financial system against climate change (Tooze 2019). Basically, climate change is causing more unpredictable, frequent, and extreme weather events, damaging property and disrupting trade, while policies to abate emissions and favor green technology have the potential to trigger sharp falls in asset prices of several industries. Providers of financial products cannot easily discharge such risks from their portfolios, an action that could potentially destabilize the entire financial system. At the same time, when—if—humanity eventually decides to stop using fossil fuels, the oil industry, one of the most heavily capitalized industries, could collapse along with demand for its products if the Big Oil is unprepared. In a sense, it is a double-edged sword: if the climate crisis hits the financial world too violently, it could produce a domino effect, shattering Big Oil. But in the reverse scenario, the financial system could also be disrupted by the destabilization of an industry unprepared for too accelerated a shift to a low-carbon world.

The other side of the double-edged sword fundamentally relates to the problem of *stranded assets*, which are fossil fuel supply and generation resources that, prior to the end of their economic life, can no longer produce returns largely as a result of changes associated with the transition to a low-carbon economy. Stranded assets are the core element of the so-called *carbon bubble* because when the reserves of oil and gas companies are deemed environmentally unsustainable, investing in the company implies relying on assets that are unusable and will, at some point, be written off. Currently, the price of these traded companies' shares is calculated on the assumption that all their reserves will be consumed, so the true costs of carbon dioxide in intensifying climate change is not taken into account in a company's stock market valuation. In fact, the oil industry risks $2.2 trillion in stranded assets in a low-carbon world. ExxonMobil is the most exposed to stranded assets, with more than 90 percent of potential capital expenditure up to 2030 failing to comply with the International Energy Agency's 1.6°C pathway, while Shell's risk is 70 percent, TotalEnergies's is 67 percent, Chevron's is 60 percent, BP's is 57 percent, and ENI's is 55 percent (Mace 2019). With regard to NOCs, experts believe that most of Venezuela's carbon-heavy blends of crude will remain stranded in the ground (Stott 2020).

Either way, the financial world must brace itself for climate change, as Kristalina Georgieva, chair and managing director of the International Monetary Fund, forcefully stresses: "Climate change is an existential threat. It is a risk that we all have to take very seriously because from the perspective of an institution that deals with economic matters, it can push back development" (Elliott 2019).

But the financial world has thus far not risen sufficiently to the challenge. While central banks and international financial institutions are under increasing pressure to engage in aiding socioeconomic systems to fight climate change, it is much less clear what their role in avoiding/preventing/coping with a climate-induced global financial crisis should be. Central banks are, for instance, urged to implement a *"green quantitative easing"* program that involves the purchase of green corporate bonds (Dafermos, Nikolaidi, and Galanis 2018), whereas the Bank of England openly acknowledges that global capital markets are financing projects likely to produce a 4°C temperature rise (Partington 2019). A global network of sixty climate organizations, led by Rainforest Action Network (RAN), issued *"Principles for Paris-Aligned Financial Institutions: Climate Impact, Fossil Fuels and Deforestation,"* which offers a timely road map for the decarbonization of the finance sector to align with the Paris Agreement (RAN et al. 2020).

The Network for Greening the Financial System, a global coalition of central banks and supervisory authorities advocating a more sustainable financial system, has urged its members to collect better data to gauge the extent of the climate risk and to advance sustainability within their own portfolios. For instance, Christine Lagarde announced that climate change would be firmly put on the European Central Bank's agenda, stating at a confirmation hearing before a European Parliament committee in Brussels that "climate change is one of the most pressing global challenges facing society today. My personal view is that any institution has to actually have climate change risk and protection of the environment at the core of their understanding of their mission" (Alderman 2019). This claim is indeed reinforced by the backing of European Union finance ministers, who urge closing the tap to funding for fossil fuels altogether (Guarascio 2019), followed by the trailblazing decision of the European Investment Bank reported above and by commercial banks increasingly distancing themselves from fossil fuel investments.

The ultimate objective of these initiatives is to avoid a sudden collapse of asset prices. And here the double-edged sword problem raises its head again, and it is a spectacular sword of Damocles in both case scenarios: a financial crisis could seriously undermine Big Oil's stability, but given the still huge overall market capitalization of oil and gas companies, their financial contraction should be careful and gradual to avoid the spectacle of witnessing the crash of the entire financial system. These reasons are in addition to those offered in chapter 6 of why a *gradual transition* scenario, whereby the phasing out of the industry's operations and products should proceed progressively, is preferable to an abrupt one, which would see an immediate dissolution of fossil fuel–related activities and thus of the industry as a whole.

According to BlackRock, the world's largest asset management firm, the harm caused by climate change will disrupt the US financial system, provoking losses of almost $4 trillion, approximately one-fifth of US GDP, an astonishing amount (BlackRock Investment Institute 2019). Put that on a global scale, as a study by Ricke and colleagues (2018) did, and the estimates are staggering, more than $16 trillion.

Economic calamities aside, even worse is the toll of climate change in terms of human suffering and environmental degradation. Climate-related disasters rose from 3,656 events in 1980–1999 to 6,681 in 2000–2019 (UNDDR and CRED 2020). In the summer of 2018 in Japan, ninety-six people were killed across the country by a heat wave. The wildfires that plagued northern California in October 2017 killed at least forty-four people, while those of the summer of 2020 caused the deaths of thirty-one, with each event provoking more than $10 billion in damages. In August and September 2017, widespread flooding during Hurricane Harvey caused at least $125 billion in damages in the Houston area and contributed to ninety-three deaths. Again in 2017, Hurricane Irma damaged $50 billion worth of property in Florida, while Hurricane Maria caused $90 billion in damages in Puerto Rico, with at least sixty people dying as a direct result of the storm. The floods that affected several European countries in the summer of 2021 killed at least two hundred forty-two people, mostly in Germany.

Scientists argue that human-induced climate change is increasing the frequency and intensity of several extreme weather events such as heat and cold waves, drought, extreme precipitations (IPCC 2021a), tornadoes and hurricanes (Trenberth et al. 2018), and the frequency and size of wildfires in much of the United States, particularly in forested areas such as the ones that ravaged California and Oregon (Abatzoglou and Williams 2016) and on the other side of the globe in Australia (Doherty 2019). The last United Nations World Meteorological Organization's *Statement on the State of the Global Climate* says that during the first half of 2020, 9.8 million displacements, largely due to hydrometeorological hazards and disasters, were recorded (WMO 2021). The CEO of Chubb, the world's biggest publicly traded property insurer, claimed that 2018 was a year of "*biblical*" catastrophes caused by climate change, with global economic losses amounting to $160 billion (Chubb 2019). In sum, a "ghastly future" of mass extinction, declining health, and climate-disruption upheavals threatens human survival (Bradshaw et al. 2021).

All these extreme events associated with climate change as well as nonextreme events, such as diminishing crop yields, insurance claims, and lower productivity, impose a cost on society, both monetary and otherwise. More generally, the increasing economic burden associated with climate change and the prospect of greater liabilities to come have brought into focus a question that has been willfully obscured in the climate change debate: Who should bear the cost of the harm—the suffering, destruction, and death—caused by human-driven climate change? Is it taxpayers—for instance, through state disaster relief funds—or affected individuals, families, and private businesses or those agents that have somehow contributed to it?

So far it is mostly states, largely through taxpayers' money, that have pledged to or do fund action favoring their own citizens or other countries harmed by climate events. But there are other agents that have played a major role in the accumulation of such costs and that, accordingly, should bear some of the brunt for financially redressing climate impacts: oil and gas companies. The analysis carried out suggests that Big Oil has a duty of reparation to sustain, and this chapter investigates how the industry can operationalize and implement this duty.

THE MORAL PRIORITY OF THE DUTY OF REPARATION
AND THE UNAVOIDABILITY OF LEGAL ACTIONS

The core moral argument of this book is founded on Big Oil's violation of the no-harm principle, which assigns the companies that make up the industry a positive moral responsibility in the form of a duty of reparation and of decarbonization. While there is much greater scientific analysis, emphasis, and rhetoric on the latter—especially embedded in the narratives of the low-carbon transition and sustainability—the duty of reparation is often overlooked and even disputed, let alone scrutinized. In fact, its investigation is possibly the book's most original contribution, given that no other analysis of the oil industry with regard to the climate crisis has attempted to comprehensively justify and shape this duty's theoretical bases and to operationalize and implement it.

Given the main moral thrust of the book, the duty of reparation is, in fact, crucial. In strict moral terms, there is even a priority for reparation over decarbonization, because the former supports those already harmed, whereas the latter is aimed at reducing future harm.

The moral priority of reparation does not mean, of course, that decarbonization is not vital; indeed, it is of the utmost importance for the planet, given both the size of the industry and its emphasized central role in shaping and maintaining global socioeconomic systems organized around fossil fuels. The point is simply that according to the theoretical analysis of part II, supporting those already harmed (duty of reparation) is a somewhat greater moral imperative than reducing future harm (duty of decarbonization).

In principle, the modification in the *health status* that Big Oil would undergo while trying to accomplish the two duties should not be of concern to this book: from an exclusively moral perspective, oil and gas companies could either cease operating immediately or transition to distributing low/zero-carbon–intensive products and clean their processes while satisfying their financial rectification requirements over a reasonable, scientifically sound period of time. In both cases, they would abide by their duty of decarbonization. In fact, this should rather be called a duty of *aspirational full decarbonization*. This duty should be understood as requiring that sooner or later—within the timeframe that current scientific evidence concedes—Big Oil, by eliminating carbon-based products and processes, becomes *Big Green Energy*, that is, if oil and gas companies wish to continue operating. In theoretical terms, oil companies are ultimately behind the decision of whether or not they want to cease operations immediately or phase out fossil fuels gradually and, in the latter instance, how to do so.

However, as remarked in chapter 6, on empirical grounds, trade-offs between the duties of reparation and decarbonization are not only significant but are also inevitable. Scenarios of decarbonization imposed from above that are too draconian would prevent oil companies from honoring their duty of reparation and deprive them of the initial financial resources needed for investing in low-carbon projects in view of a *full decarbonization*. Therefore, in the pragmatic perspective of this chapter, the industry should avoid pursuing excessively rapid decarbonization that would expose it to concrete risks of an untimely termination of its activities and to the consequent impossibility of fulfilling the duty of reparation. To satisfy this latter duty, these companies should draft plans to phase out fossil fuels over a period of time compatible with their capacity and circumstances that take account of the window of opportunity the climate crisis allows in a way that protects more vulnerable stakeholders and initially generates enough turnover to fund lower-carbon investment.

Given the goals of the duty of reparation that, as said, reflect an explicit set of societal expectations and interests divergent from those of oil and gas companies, the processes to induce them to achieve this duty are challenging and not self-propulsive, as clarified in part II. It will be no mean feat to induce any oil company, be it an international oil company (IOC) or a national oil company (NOC), to relinquish a sizable share of its income to satisfy the moral requirements demanded by the quota of harm caused by its activities. To further clarify, it is indeed possible and in many respects even likely that in the (near) future, oil and gas companies, finding themselves with their backs against the wall and cornered by an increasingly pugnacious public opinion, decide to disburse funds to financially rectify some climate harm. By the same token, it is possible that these companies' moral responsibility provides the necessary theoretical trigger to hold them legally liable, with courts ordering them to undertake appropriate remedies for their faulty actions.

The proliferation of responsibility-based climate change lawsuits brings to the fore how both authorities and organized agents of destabilization are using legal procedures to determine how carbon emissions and their impacts should be addressed. Political authorities are seeking to hold the industry liable for the harm produced by shifting part of the cost of protection to the companies: this is indeed a form of financial rectification of the harm done. However, given that this form of harm rectification is unrelated to morally grounded considerations and the consequent partiality of the scope of liability-based initiatives, an approach such as this would not comply with the moral requirements of the duty of reparation; this duty, in fact, calls for an oil company to rectify its morally justified portion of the harm done. (This crucial point on the specification of harm consistent with the moral provisions of the duty of reparation is addressed in the ensuing section.)

Despite such voluntary and legally driven initiatives, in all likelihood Big Oil will not spontaneously comply fully with its duty of reparation, as repeatedly stressed. Oil and gas companies—collective entities that pursue specific industrial objectives on behalf of their shareholders and owners—generally have no mandate or incentive to disburse money to provide services outside their industrial scope, that is, to financially rectify the harm inflicted on the planet and humanity overall.

It could be argued that the duty of decarbonization is faced with an analogous situation. In the end, the industry must bear some costs if it follows the path of a less carbon-intensive business model that aims at providing a (global) public good—climate stability—that is still not one of its industrial goals despite the rhetoric. However, the emergence of a norm condemning the use of fossil fuels as harmful products prompting significant anti–fossil fuel initiatives, coupled with the industrial opportunities given by technological breakthroughs in already established renewables, as well as the development at scale of carbon-negative approaches may induce the oil industry to develop a prompt (if partial) transition to lower/zero-carbon products and processes. These external pressures and possibilities can therefore favor an endogenous reorientation of the oil industry's business model coherent with the duty of decarbonization, as many oil and gas companies indeed increasingly claim.

On the contrary, the duty of reparation lacks, so to speak, an *opportunity/cobenefit side*; this deficiency, coupled with the lack of a mandate and incentives for disbursing funds to redress the harm suffered by external, unrelated subjects would almost certainly limit voluntary actions to fulfill the duty of reparation.

It can thus be surmised that exogenous action is unavoidable. Such action, in turn, needs to be enabled, by and large, by the agents of destabilization defined in chapter 8 as *primary*. Their role and objective is to prepare the ground for inducing society to accept and acknowledge the future inadmissibility of fossil fuels. On this fertile terrain, political authorities at various levels will feel more empowered to introduce the opportune legal provisions necessary to ensure that companies financially rectify fossil fuel–related harm. No legal provision can put a precise monetary figure on the harm done, as the following section explains; the law, however, can approximate, perhaps through iterative recalibrations, such objectives. At any rate, the most effective way for operationalizing the duty of reparation first and foremost requires political authorities to adopt legal provisions that oblige financial rectifications on Big Oil.

In many respects the example of the tobacco industry can be indicative for the oil industry (Shulman 2012), not only with regard to denial. The legal path in that case was pursued through the 1998 Tobacco Master Settlement Agreement. The defendants were originally the four largest American tobacco companies (Philip Morris, R. J. Reynolds, Brown & Williamson, and Lorillard, the *original participating manufacturers*, referred to as the *"majors"*), with the case being brought by the attorneys general of forty-six US states on various grounds; in particular, the claim was for damages to cover annual costs incurred by the states to cover the medical treatment of people with smoking-induced illnesses. The money raised also funded an antismoking advocacy group, called the American Legacy Foundation and later renamed Truth Initiative. Without entering into the technical details of the judicial case—largely outside the scope of the current argument—it is quite clear that to compel Big Oil to meet its duty of reparation, a similar strictly enforced legal provision is crucial.

For the legal option to gain momentum and especially have a global knock-on effect across sovereign states, a declaration or a similar arrangement could be conceived and agreed upon by governments and the relevant international governmental and nongovernmental organizations. Such an agreement should outline nonbinding principles to formulate how funds are to be disbursed when the oil companies comply with their duty of reparation.

An initiative of this sort would, for instance, be in line with the 1998 Washington Conference on Holocaust Era Assets, which endorsed a set of principles—known as *the Washington Principles on Nazi-Confiscated Art* and sometimes referred to as the *Washington Declaration*—to promote the restitution of art confiscated from the Jewish population in Germany by the Nazi regime before and during World War II and to help the heirs of Jewish collectors recover Nazi-looted art. Despite the duty imposed by the Washington Declaration not being in the form of *disgorgement* but rather of *restitution*, in terms of the specifications of the forms of the duty of reparation outlined in chapter 6, what counts here is the underlying logic of the approach. In other words, oil and gas companies can be induced to fulfill their duty of reparation through a similar concerted international effort, where states and international governmental and nongovernmental organizations agree to the terms for disgorging funds, possibly in accordance with the companies themselves.

However, the operationalization and implementation of the duty of reparation by the individual industry agents greatly depends on the politics, the societal context, and the efficacy of agents of destabilization of the home states, as shown later.

THE MORALLY PERTINENT HARM TO BE RECTIFIED:
A FOREMOST OPERATIONAL ISSUE

This section aims to define in empirical terms the *morally pertinent* harm that political authorities, through laws and regulations, should compel Big Oil to financially rectify. There is an emerging field of investigation focused on agents' accountability for the social losses and damages stemming from climate change that should be pursued, as evinced by the previously mentioned proliferation of lawsuits against fossil fuel companies and negotiations for losses and damages under the United Nations Framework Convention on Climate Change (UNFCCC) (Williams 2020).

Financial rectifications required by the duty of reparation first involve clarification of which impacts would have naturally occurred versus those attributable to anthropogenic climate change. Obviously, Big Oil cannot be held morally responsible for any harm falling into the former category.

Generally speaking, the causal chain that goes from human influence on climate change to distinct impacts on human, socioeconomic, and natural systems can be clarified through different kinds of approaches of the previously mentioned *attribution science* (Marjanac et al. 2017), usually through a risk-based system that addresses this point probabilistically or a storytelling approach that inspects the role of the various

factors contributing to the event and decides its attributability determin-
stically (Shepherd 2016).

A first step, known as *detection of change,* requires proving that a particular variable has changed in a statistically significant way. The second
step, *factor attribution,* involves identifying the possible causative factors
to determine the role of one or more drivers with respect to the detected
change and the consequent harm. Eventually, *source attribution* seeks to
ascribe any change to specific agents (Burger, Horton, and Wentz 2020).

In an operational perspective, attribution science with regard to Big
Oil first entails attributing impacts to climate change and then attributing harm to climate-related impacts. In the first case in point, the aim
of the rectificatory action can be narrow, focusing only on impacts reliably attributable to anthropogenic climate change, or it can be broad,
focusing on all impacts associated with climate variability. It seems quite
obvious that the current analysis favors the narrow focus of rectification
of impacts, ones that can dependably be attributed to anthropogenic climate change so as to provide a credible normative reference for rectifications by clarifying the link between the industry's responsibility and its
contribution to the problem. The second point is more complex, as it
requires evidence for attributing harm to climate-related events. On the
one hand, as mentioned, attribution science offers new evidence about
the chain of causality between fossil fuel use and climate change-related
impacts (James et al. 2019; Burger et al. 2020).

It remains, however, unclear how climate attribution research can in
its current level of development inform climate policy and justice debates
with particular regard to rectification of the harm done (Huggel et al.
2016; James et al. 2019). Climate policy and climate negotiations are yet
to clarify what type of evidence would be required for claims of financial
rectification even in cases when further steps toward liability, such as the
legal duty to pay for remedying the negative effects of climate change,
are invoked.

To unravel these complexities, it is worth noting that by and large, it
is events that cause harm whose adaptation is impossible. Slow-onset

(anthropogenic) climate change in general does not count toward harm in the form intended here, as it is assumed that populations have the capacity to adapt to its gradual rate of occurrence. However, those slow onset climate impacts that cannot be adapted, such as global sea level rise, do. Similarly, extreme events can be evinced as referring to fast paced climate change, to which adaptation is almost impossible and that therefore produces inevitable harm to humanity and the planet. This seems a sensible and widely acceptable assumption: its rationale is that in the last decade, scientists are increasingly adept at identifying and quantifying how far anthropogenic climate change increases the frequency and intensity of a variety of extreme events (Schiermeir 2018; UCS 2018b; Pfrommer et al. 2019). At the same time, attribution science related to slow climate change and to the influences on harm caused by the interaction of climate change with other drivers of risk, including socioeconomic ones, is rapidly advancing (James et al. 2019).

For instance, the US National Academy of Sciences states that it is now possible to make and defend quantitative statements about the extent to which anthropogenic climate change has influenced either the magnitude or the probability of the occurrence of weather events (NAS 2016). Similarly, the *Bulletin of the American Meteorological Society*, which has published a yearly report on climate detection and attribution annually since 2012 (for the last available report, see Herring et al. 2021), reiterates that scientific advancement makes it possible to detect the *fingerprints* of climate change on any given event with some precision. There is strong evidence that extreme heat waves, coastal flooding resulting from storm surges and regular high-tide events, and severe precipitation, including hurricane downpours, have a strong causal link with climate change whose contributions to these events are, in fact, most certainly identifiable. For instance, as said above, the record-breaking July 2019 heat wave in Western Europe was intensified by human-induced climate change, and its observed magnitudes would have been extremely unlikely without it. The science is, however, less conclusive in relation to tornadoes, thunderstorms, and some types of droughts, whereas there is growing evidence for wildfires (Knutson 2017). Furthermore, in the last few years, science has been able to discern the role of climate change on *individual* extreme events, such as heat waves, droughts, extreme cold snaps, and the intensity of hurricanes (Schiermeir 2018; UCS 2018b).

Source attribution goes even further in trying to identify and ascribe climate impacts to specific sources; a *source* could be a particular agent (e.g., a country or a company), a sector, or an activity (Burger et al. 2020). Source attribution would make it possible to allocate a pertinent part of anthropogenic climate harm to individual oil and gas companies. This attribution is based on the proportional contribution of the company's fossil fuels to changes in global atmospheric composition, the extrapolation of the proportional contribution to localized events, and the identification of the actual harm caused by those impacts (Burger and Wenz 2018). In other words, it seems that a sound causative chain linked from anthropogenic climate change to harm and the consequent monetary costs and then to emitters—e.g., Big Oil—is now possible.

This very approach is being adopted in harm-based lawsuits against Big Oil and is being widely touted as the new frontier in tackling climate change (Leber 2019; Stuart-Smith et al. 2021). For instance, three California communities—Marin and San Mateo Counties and the city of Imperial Beach—are suing thirty-seven oil, gas, and coal companies for climate-related damages to public property such as beaches and parks, citing the possibility that some residents of these communities will lose their property and be displaced due to extreme weather events (Shulman 2017). Maryland's capital city Annapolis—which in 2020 experienced 65 days of flooding and could experience almost 200 days of annual flooding by 2030 and 350 days by 2040—has filed suit against twenty-six oil and gas companies, with city officials citing them for responsibility for rising sea levels; the defendants include BP, Chevron, ExxonMobil, and Shell as well as the American Petroleum Institute (Hasemyer 2021).

The pathbreaking studies by Ekwurzel and colleagues (2017) and Licker and colleagues (2019) show that scientific evidence can help apportion responsibility and duties for climate damages among carbon producers. As stressed in chapter 1, these studies quantify sea level rises, the increase in global surface temperatures, and the acidification of the world's oceans that can be traced to the emissions of specific companies. For example, more than 6 percent of the rise in global sea level is caused by emissions associated with three large investor-owned contributors, BP, Chevron, and ExxonMobil, and emissions traced to the twenty companies named in the California communities' lawsuits contributed to 10 percent of global sea level rise over the same period (Frumhoff and Allen 2017). Furthermore, NASA provided an approach for calculating the individual driving forces of recent climate change through direct satellite observations (Kramer et al. 2021); while a high-tech independent effort has been able to track carbon dioxide and methane emissions from specific countries, industrial facilities, and power plants through satellite data, machine learning, and artificial intelligence (Freedman 2021).

Given this scientific background, attribution science certainly has important political and moral implications (Mechler and Schinko 2016)—which are complex and often difficult to unravel (Burger and Wenz 2018)—and should ensure that it uses a pertinent notion of vulnerability of impacted subjects (Stone et al. 2021). Attributing specific harm to carbon emissions can imply responsibility and duties for emitters, including countries, regions, sectors, individuals, and, indeed, the very companies that make up the oil industry. Of course, attribution science is not sufficient and does not aim to establish emitters' moral responsibility, which is a multifaceted issue that extends far beyond climate science (Wallimann-Helmer et al. 2019) and involves investigating agents' intentions, knowledge, voluntariness, and control, as shown in chapter 2. In other words, determining who should bear the burden of climate harm and what such harm is are so far largely a social and political question.

As said, in principle, the harm that should be rectified by Big Oil should be the anthropogenic fraction of the overall harm, attributable to its cumulative emissions based on scientific evidence and mostly caused by extreme and slow-onset events that cannot be adapted to, like the steady rise of the oceans. This is, indeed, a very specific definition of the harm to be rectified; given its sound basis, it should be quite uncontro-versial and hence widely acknowledgeable. However, in practical terms, complete and indisputable evidentiary bases for attributing the damages caused by emissions are still lacking. These approaches nonetheless pro-vide a practical first step in the recognition of responsibilities and a pos-sible *reconciliation* process between those who caused climate harm and those being impacted by it (Huggel et al. 2016). Therefore, this framing makes it possible to indicate, in theory, the morally pertinent harm that Big Oil should financially rectify.

Unfortunately, though, in empirical terms, even the most advanced attribution science cannot yet provide such detailed, company-specific information and is unlikely to do so for some years to come, especially as the majority of the industry's emissions are of the scope 3 kind, that is, they originate from the downstream combustion of the products they have distributed throughout the global economic system, and their iden-tification and calculation are more elusive. Therefore, while awaiting attribution science developments in estimating an indisputable mone-tary value of the morally pertinent harm caused, there is a pressing need to arrive at an empirically feasible bridging solution to estimate Big Oil's duty of reparation, one that is nonetheless solid.

In other words, given the current climate emergency and the critical role played in it by Big Oil, a measurable approximation of the fraction of climate harm that an oil company is morally compelled to financially rectify is needed. One possible option—based on the social cost of carbon (SCC) associated with the industry's contribution in terms of emissions—will be sketched out below to provide a model of the implementation of the duty of reparation for the top twenty oil and gas companies.

FURTHER OPERATIONAL ASPECTS OF THE DUTY OF REPARATION

After the all-important clarifications needed to frame theoretically the morally pertinent harm that the duty of reparation should actually address, its fully fledged operationalization needs further specifications.

In particular—consistent with the more pragmatically oriented specifications of the duty of reparation outlined in chapter 6—details are needed of the form and concrete means that rectificatory action should take to effectively meet duty recipients' requirements; another obvious specification requires identifying who these recipients are and to what extent they are entitled to reparations.

The financial rectification demanded of Big Oil by the duty of reparation could be accomplished through a global fund that could be named the *Fund for Oil Rectification* (FOR). Binding international legal initiatives should establish the FOR as an independent, global legal entity and investment vehicle to help mobilize and supervise the collection and allocation of financial resources disgorged by Big Oil.

Ideally, the FOR should be administered by trustees selected among members of civil society, governmental and nongovernmental organizations working on climate change, science and education, environmental issues, justice, peace and security, development, international law, financial matters, and scientific communities. They will be subject to a mechanism—examined later in this section—for monitoring the FOR's activities to ensure efficiency and to avoid the possibility of corruption or malfeasance. These trustees or fund administrators should have long-term appointments, as the regular interruption of mandates can distract from longitudinal goals to focus efforts on short-term solutions. As an institution through which staggering sums of money will pass, the FOR must be prepared to be subjected to rigorous public and media scrutiny to ensure it is *above suspicion* in the assignation of funding. Thus, the FOR must have financial disclosure policies, protocols to ensure third-party accountability, whistleblower protection, and any other instrument necessary to maintain institutional integrity and safeguard it from charges of corruption.

The FOR administrators will have a clear mandate set in stone to enact its funding mechanisms while fully safeguarding the climate system for the benefit of current and future generations. A financial mechanism of this kind would facilitate strategic focus, rigorous project management, solid monitoring and evaluation, and high levels of transparency. Its structure should be similar to that of a *sinking fund*, whose entire principal and investment income is disbursed over a fairly long period—in the case of Big Oil, a starting point could be to set the terms over a thirty-year period—until it is exhausted and thus reduced to zero. Its capitalization and resource mobilization strategies are exclusively dependent on money disgorged by oil and gas companies.

Any local source of emissions concurs with the global increase of the concentration of greenhouse gases in the atmosphere, so given the undifferentiated global origin of Big Oil's contribution to climate change deriving mostly from the global span of its processes and products, the FOR should be truly global in its scope. It should not take into account any regional/national/local/sectoral distinctions in terms of financial replenishment or the disbursement of its funding.

The FOR should involve three units: a core unit that funds the people most vulnerable to climate change (the *harm* unit) and two subsidiary units, one aimed at promoting the low-carbon transition by addressing the social burden of the duty of decarbonization and supporting clean technologies, projects, and so on (the *transition* unit), and the other at assisting the direct *victims* of Big Oil's decarbonization, that is, displaced workers and frontline communities (the *workers and communities* unit), as shown in figure 9.1. Thus, the transition and the workers and communities units would support socially and environmentally beneficial actions, initiatives, and projects in favor of the low-carbon transition as well as workers and communities that rely on the fossil fuel value chain (Healy and Barry 2017).

The FOR's operational facets of the replenishing and disbursing procedures should be established in socially agreed-upon ways decided by all the relevant stakeholders, taking similar existing financial mechanisms—such as the UNFCCC Green Climate Fund and, perhaps, the future European Union Just Transition Fund—as possible models. Accordingly, the remainder of this section limits its scope to the clarification of the main features of its three units in terms of access of duty recipients and will outline a possible mechanism for monitoring the fund's activity.

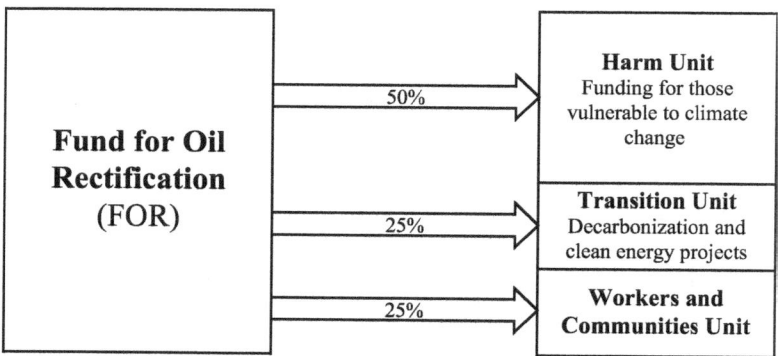

Figure 9.1
Proposed Fund for Oil Rectification (FOR).
Source: Author.

The FOR's core unit should have the sole aim of rectifying the harm endured by those most socially vulnerable to climate change at a global level. However, given the factual impossibility of addressing all morally pertinent harm suffered, only a fraction of it can hope to be redressed, as described in the ensuing model for the implementation of Big Oil disgorgements. At any rate, this FOR unit—given its stringent objective of enacting the requirements of the duty of reparation and its moral salience as well as the urgency of climate harm for more vulnerable peoples—should be the principal one. Ideally, half the disgorged funds should be channeled into the harm unit, the other half being divided equally between the subsidiary transition and workers and communities units. Access to funding from the FOR harm unit should be inversely proportional to the social vulnerability level of the recipients.

With regard to the relatively smaller transition unit, disbursed funds should adopt an approach that privileges the most effective action, initiatives, and projects to support the low-carbon transition, ensuring that part of its plan includes funding to decommission the thousands of oil and gas wells located worldwide to minimize their possible reuse. In the past, the cost of closing these disused onshore and offshore sites, involving the removal of topside facilities and the capping of the well itself, was always covered by the future incoming profits of the owner companies, but in a scenario aiming squarely at low-carbon technologies and keeping fossil fuels in the ground, the logic of this revenue flow no longer holds. Decommissioning costs in the United Kingdom alone were estimated to be £1.1 ($1.48) billion in 2020, according to the country's association of the offshore oil and gas industry (Oil & Gas UK 2020).

As for the workers and communities unit, all displaced workers and communities should be entitled to a share of the fund parametrized to their former income or welfare, perhaps capped at the level of the country's median income. Indeed, protecting displaced workers and the communities making their living through fossil fuel production—this FOR unit can be employed, for instance, for income support, retraining, the provisions of other public services to the community, and remediating contaminated sites—is crucial for gaining the acceptability and feasibility of anti–fossil fuel initiatives and should be undertaken in close collaboration with national and subnational governing bodies. For instance, in the 2020 unsuccessful race for his candidacy to become the US Democratic presidential nominee, Bernie Sanders argued, in relation to his climate plan, that "this plan will prioritize the fossil fuel workers who have powered our economy for more than a century and who have too often been neglected by corporations and politicians. We will guarantee five years of a worker's current salary, housing assistance, job training, health care, pension support, and priority job placement for any displaced worker, as well as early retirement support for those who choose it or can no longer work" (Sanders 2019).

It is worth stressing that according to the quantitative indications provided as a model in the next section, these two subsidiary units—transition and workers and communities—would be endowed with roughly $600 billion over the period 2020–2050. The Just Transition Mechanism of the European Green Deal Investment Plan—which by and large has the same objectives as the workers and communities and transition units of the FOR—is estimated to mobilize €143 ($161.3) billion over a ten-year period to support European Union regions most affected by the transition

(European Commission 2020).[1] These two financial sources for support-ing the low-carbon transition, despite the different territorial scope, have therefore roughly the same order of magnitude.

Finally, given the centrality and sensitivity of the entire reparation process, it is vital that the functioning and effectiveness of the FOR be subject to regular checks and a thorough critical review. To facili-tate this monitoring process, the FOR should include some overarch-ing and cross-cutting calibration and adjustment mechanisms involving epistemic qualities that provide the evidence required to achieve its ulti-mate goal, that of supporting the most vulnerable subjects in dealing with climate-related harm. Among these epistemic qualities, two of the most prominent in relation to the nature and objectives of the FOR are *accountability*, that is, the demand that the fund abide by certain codes of conduct and the prospect of judging whether it actually conforms to that conduct; and *transparency*, that is, the possibility of monitoring the running of the fund so as to avoid malfeasance. These qualities would allow the consistency of the FOR's conduct and mission to be under-stood and evaluated.

AN EXEMPLARY IMPLEMENTATION OF THE DUTY OF REPARATION

In view of the implementation of the duty of reparation through the FOR, the quantitative contribution to harm by each specific oil and gas company must first be established. The figures for the twenty top oil and gas companies in terms of scope 1 + 3 cumulative greenhouse gas emis-sions for the period 1990–2015 are reported in table 9.1: 1990 is, in fact, commonly agreed upon as the year of widespread acknowledgment of climate change, although this is a very generous concession to Big Oil in light of the evidence supporting morally relevant fact A, awareness, reported in chapter 2. The figures of table 9.1 are elaborated from the Carbon Majors Database—2017 Dataset Release (CDP 2017); there, how-ever, oil and gas companies' emissions were calculated for the period 1988–2015. Here, emissions are, as said, calculated from 1990 to 2015,

Table 9.1

Top 20 oil and gas companies' scope $1+3$ greenhouse gas emissions 1990–2015, GtCO$_2$e and percentage of contribution to harm

Oil and Gas Company	Emissions	Harm (%)
Saudi Aramco	36.1	14.8%
Gazprom (Russia)	31.3	12.9%
National Iranian Oil	18.2	7.5%
ExxonMobil (USA)	15.8	6.5%
Pemex (Mexico)	14.9	6.1%
Shell (UK/Netherlands)	13.3	5.5%
CNPC/PetroChina	12.4	5.1%
BP (UK)	12.3	5.0%
Chevron (USA)	10.5	4.3%
PDVSA (Venezuela)	9.8	4.0%
Abu Dhabi National Oil—ADNOC	9.6	3.9%
Sonatrach (Algeria)	8.0	3.3%
Kuwait Petroleum	8.0	3.3%
TotalEnergies (France)	7.6	3.1%
ConocoPhillips (USA)	6.7	2.7%
Petrobras (Brazil)	6.1	2.5%
Lukoil (Russia)	6.0	2.4%
Nigerian National Petroleum Corp	5.8	2.4%
Petronas (Malaysia)	5.5	2.3%
Rosneft (Russia)	5.2	2.2%
Top 20 O&G Companies	**243.1**	**100%**

Source: Elaboration from the Carbon Majors Database—2017 Dataset Release (CDP 2017). According to the Greenhouse Gas Protocol of the World Resources Institute (WRI n.d.), scope 1 emissions are to direct oil and gas combustions; scope 3 emissions originate from the downstream combustion (for energy and nonenergy purposes) of oil and gas distributed within the global economic system.

based on the simplified assumption of annual contributions being constant over the entire 1988–2015 period considered by the database.

It should be pointed out that the figures provided in table 9.1 and, more generally, all the figures relating to the implementation of the duty of reparation and decarbonization are only indicative, though in a rigorous, scientific manner, of the scale of the issue at stake. They by no means aim to confute the hard numbers that only case-specific investigations can produce but instead aim only to provide an idea of the magnitude of the involvement required of individual oil and gas companies.

According to table 9.1—in which the entire financial burden of the harm is shouldered by top twenty emitters—Saudi Aramco, the largest company with a sum total of 1990–2015 emissions of 36.1 $GtCO_2e$, can, for instance, be attributed a quota of harm equal to 14.8 percent, while Rosneft, of the companies under consideration the smallest contributor, has a quota of harm of 2.2 percent for its 5.5 $GtCO_2e$ of emissions.

Each company should be expected to contribute to the FOR an approximate amount based on its contribution in terms of emissions measured through the previously mentioned SCC. This money metric is a measure of climate harm, expressed as the dollar value of the cumulative economic impact of global warming attributed to each tonne of carbon released into the atmosphere. Currently, a credible estimate of the SCC that collates data produced from a number of expert sources is around $90 per tonne (Pindyck 2019). The estimate of the SCC is extremely controversial (Wagner et al. 2021). For instance, the interagency group established by US president Joe Biden has set the SCC at an interim value of approximately $51 per tonne, but it could reach as high as $125 (IWG 2021). Ricke and colleagues (2018) estimate a median SCC of $417 per tonne, while Tol (2019) underlines that it differs greatly between countries and tends to be highest in poor countries with large populations. At the same time, ExxonMobil is believed to internally use an SCC (the company calls it *"approximate cost of carbon,"* which is the figure it employs for estimating the risk that climate change and changing energy demands pose to its projects) of $80 per tonne of CO_2 in developed countries until 2040.

Pindyck's SCC estimates employed here are based, besides modeling, on opinions of people with research experience and expertise in climate change, policy, and its impact (climate scientists as well as economists and other social scientists). Based on this coupled modeling/survey approach, Pindyck's estimates yield an average SCC of $200 per tonne.

He further proceeded by "trimming outliers and focusing on experts who expressed a high degree of confidence in their answers" (Pindyck 2019, 140) and established a lower SCC of $80–$100 per tonne, hence the $90 SCC average employed here.

Accordingly (bearing in mind that 1 Gt equals 1 billion tonnes), the financial estimate of the harm generated by Saudi Aramco cumulative 1990–2015 emissions would be $3,248 billion, while that of Rosneft would be $472 billion (table 9.2).

Table 9.2

Top 20 oil and gas companies' harm measured through the social cost of carbon (SCC = $90 per tonne)

Oil and Gas Company	Morally Pertinent Harm ($ billion)
Saudi Aramco	3,248
Gazprom (Russia)	2,816
National Iranian Oil	1,640
ExxonMobil (USA)	1,424
Pemex (Mexico)	1,344
Shell (UK/Netherlands)	1,200
CNPC/PetroChina	1,120
BP (UK)	1,104
Chevron (USA)	944
PDVSA (Venezuela)	880
Abu Dhabi National Oil—ADNOC	864
Sonatrach (Algeria)	720
Kuwait Petroleum	720
TotalEnergies (France)	680
ConocoPhillips (USA)	600
Petrobras (Brazil)	552
Lukoil (Russia)	536
Nigerian National Petroleum Corp	520
Petronas (Malaysia)	496
Rosneft (Russia)	472
Top 20 O&G Companies	**21,880**

Source: Author.

Table 9.2 presents astonishing figures, breathtaking sums of money much greater than oil and gas companies' total assets. It is therefore materially impossible for Big Oil to cover the entirety of the morally pertinent harm caused or even substantial amounts of it when it is measured through the SCC.

Given this evidence and considering that the world's largest oil and gas companies' *exploration and production* capital expenditures (capex)—that is, the money employed for exploring and producing oil and gas—amounted to $546 billion in 2019 (Rystad Energy 2020a), it seems sensible for a feasible yet morally cogent duty of reparation to first demand that those expenditures be set to zero—over the shortest possible period of time compatible with the industrial and financial obligations of the different fixed assets—and actually turned into financial rectifications of the harm done. Additionally, no company in the industry should be permitted any further capex in fossil fuels. However for the demonstrative objective of maintaining the connection with the moral basis of the duty of reparation, its implementation should include a portion of the morally pertinent harm as estimated by the SCC in table 9.2. To strike an empirically constrained and yet morally significant balance between the ambition of a fully-fledged satisfaction of the duty of reparation and the current reality of Big Oil's financial capacity, a drastic resizing of such amounts should be undertaken, taking into account the trade-off with the duty of decarbonization.

To this end, a simple reference scheme for rectification in line with the duty of reparation can be based on the map of Big Oil in terms of possible requirements demanded by its duties of reparation and decarbonization developed in chapter 7. That map envisages three groups of oil and gas companies with differing levels of requirements: high requirement (HR companies: BP, Chevron, CNPC/PetroChina, ExxonMobil, Gazprom, Shell, and Saudi Aramco), medium requirement (MR companies: ADNOC, ConocoPhillips, Kuwait Petroleum, National Iranian Oil, PDVSA, Pemex, Petrobras, Petronas, Rosneft, and TotalEnergies), and low requirement (LR companies: Lukoil, Nigerian National Petroleum, and

Sonatrach). As a result, the burden of the financial rectification should be distributed accordingly to the three Big Oil groups (HR, MR, and LR). HR companies are expected to contribute the most: they would, for instance, cover 3 percent of the harm associated with the SCC of percentage contribution in table 9.1. This amount would then be ramped up by 0.1 percent—based in each time period on the companies' contributions calculated in the same table and therefore, for the sake of simplicity, not compounded—each year until 2050. The maximum sum contributed in 2050 would therefore amount to 6 percent of the SCC associated with their emissions. The MR group of oil and gas companies would be required to disgorge funds according to the same procedure followed by HR companies but would start disgorging an amount equal to 2 percent of their SCC; through the subsequent 0.1 percent increases over the following thirty years, they would reach 5 percent of their SCC in 2050. LR companies would start from 1 percent and arrive at 4 percent of their SCC in 2050.

The reason for the choice—admittedly subjective—of these particular percentages is twofold. On the one hand, they are realistic figures in relation to the economic and financial size of oil and gas companies, while at the same time they are able to replenish the FOR with a substantial amount of financial resources. On the other hand, they respect the different levels of requirement demanded by the duties of oil and gas companies and also respect the trade-off between the duties; that is, they indicate a path to reparation gradual enough not to disrupt an adequate operationalization and implementation of the duty of decarbonization.

Table 9.3 provides an indicative overview of a prudent implementation of financial rectification through the FOR of a portion of the morally pertinent harm for the different companies, grouped according to their level of requirements. Any ongoing capex already committed should be added to these substantial sums, as clarified above. Altogether, these figures can be considered benchmarks, aspirational objectives that the largest oil and gas companies should pursue in ideal specific situations and contexts.

Table 9.3

Top 20 oil and gas companies' 2020–2050 disgorgement (all figures in $ billion)

Level of Commitment	Oil and Gas Company	Morally Pertinent Harm	Initial Disgorgement (2020)	Subsequent Annual Disgorgements (2021–2050)	Total Cumulated Disgorgement (2050)
High requirement (initial disgorgement 3%)	BP (UK)	1,104	33.1	1.1	66.2
	Chevron (USA)	944	28.3	0.9	56.6
	CNPC/PetroChina	1,120	33.6	1.1	67.2
	ExxonMobil (USA)	1,424	42.7	1.4	85.4
	Gazprom (Russia)	2,816	84.5	2.8	169.0
	Shell (UK/Netherlands)	1,200	36.0	1.2	72.0
	Saudi Aramco	3,248	97.4	3.2	194.9
Aggregate HR					**711.4**
Medium requirement (initial disgorgement 2%)	Abu Dhabi National Oil—ADNOC	864	17.3	0.9	43.2
	ConocoPhillips (USA)	600	12.0	0.6	30.0
	Kuwait Petroleum	720	14.4	0.7	36.0
	National Iranian Oil	1,640	32.8	1.6	82.0
	PDVSA (Venezuela)	880	17.6	0.9	44.0
	Pemex (Mexico)	1,344	26.9	1.3	67.2
	Petrobras (Brazil)	552	11.0	0.6	27.6
	Petronas (Malaysia)	496	9.9	0.5	24.8
	Rosneft (Russia)	472	9.4	0.5	23.6
	TotalEnergies (France)	600	13.6	0.7	31.0
Aggregate MR					**412.4**

Low requirement (initial disgorgement 1%)	Lukoil (Russia)	536	5.4	0.5	21.4
	Nigerian National Petroleum Corp	520	5.2	0.5	20.8
	Sonatrach (Algeria)	720	7.2	0.7	28.8
	Aggregate LR				**71.0**
	Total Disgorgement				**1,194.8**

Source: Author.

The disgorgement of funds toward the three units of the FOR required of both IOCs and NOCs is evidently very significant. In order to thoroughly investigate the implementation of the duty of reparation, the following section abandons the schematic approach—useful as it may be for giving a quantitative insight into the issues at stake—followed thus far and employs a finer-grained qualitative lens that highlights some prominent political, social, and economic features related to Big Oil groups and to individual companies within them, in view of the implementation of the duty of reparation. These considerations will be strengthened by some closer insights into the possible—differentiated—role of agents of destabilization to favor the achievement of this duty.

THE ACTUAL IMPLEMENTATION OF THE DUTY OF REPARATION: POLITICS, SOCIETAL CONTEXT, AND THE POTENTIAL OF AGENTS OF DESTABILIZATION

As part II suggests, to avoid the risk of indeterminacy and increase the *real-world* influence of the duty of reparation, its implementation should be mostly a matter of political action within a given societal context where specific agents of destabilization operate. The overview of the implementation process provided below is an attempt to contextualize the duty of reparation within a realistic framework of near-term—given the urgency of the climate crisis—political evolution. To be sure, an eco-socialist future of the nationalization of industry majors would indeed make the implementation of the duty of reparation much easier. At the same time, it remains rather utopian, despite the dramatic global acceleration of climate action since the United Nations 1.5 report—at least for the modestly radical ambition of this book, which in fact rather aims to provide a *road map* for Big Oil to meet its duties deriving from its responsibility for the climate crisis in gradual, democratic, and nonviolent ways and contexts—because current world politics still seems stuck in high-carbon–intensive production and consumption patterns.

That said, chapter 7 argued basically that richer countries' (typically Western) IOCs are more likely to address their duty of reparation, whereas non-Western NOCs perhaps lean more toward waiting until there is a wider diffusion of favorable social, cultural, political, and moral factors and norms. Additionally, NOCs of more impoverished oil-exporting countries should be granted a less stringent duty of reparation—as accounted for in the categorization of their level of requirement—and given the right to development of these countries and their citizens. This can also be intended as a way to provide them with assistance for the burden they must shoulder in the process of climate stabilization and belongs to a more general right to escape poverty or disadvantage (Armstrong 2020).

Based on these considerations, to break free from the—necessarily—schematic and rigid outline of disgorgements required by the duty of reparation examined in the previous section, it is useful to cluster oil and gas companies based on their most significant features with regard to this duty as well as by the determinants that condition the implementation. To this end, a qualitative finer-grained estimate of the duty of reparation needs to include within the HR/MR/LR commitment grid the IOC versus NOC distinction as well as the influencing determinants examined in chapter 7 and, as emphasized, should attribute more weight to the *assets* objective parameter of Big Oil mapping.

It seems most opportune to create three clusters of industry giants for the implementation of the duty of reparation. The first includes HR and MR IOCs, the second includes HR NOCs, and the third includes MR and LR NOCs.

The first group of IOCs—BP, Chevron, ConocoPhillips, ExxonMobil, Shell, and TotalEnergies—should be compelled to abide by their duty of reparation on extremely short notice. It should be noted that these companies are herein assigned greatest moral responsibility given their systematic violation of the no-harm principle, as evinced by the morally relevant facts described in part I of the book. Additionally, with different degrees of involvement, they were—and some of them still are—at the forefront of the climate denial campaign that has substantially slowed down climate action. In fact, North American IOCs were, for instance,

the most aggressive in terms of climate denial. At the same time, as tes
tified by the surge of lawsuits and by the deep and effective grassroots
work carried out by primary agents of destabilization as well as by the
effectiveness of the initiatives of some operational agents, the United
States seems ready—possibly starting from the subnational and state
levels—to introduce the necessary legal instruments to establish bind
ing disgorgements, consistent with the requirements of the duty of repa
ration. US-based IOCs—Chevron, ConocoPhillips, ExxonMobil—also
seem to have the means and capacity to shoulder the financial burden
of the disgorgements indicated in table 9.3, consistent with the assets
objective parameter. All in all, the theoretical and empirical consider
ations developed in this book suggest that the duty of reparation should
be implemented with regard to US-based IOCs promptly and with
no delay.

The situation of European IOCs—BP, Shell, and TotalEnergies–is slightly
different. On the one hand, they are somewhat less morally responsible
than US ones: BP and Shell's involvement in the denial machine was
less entrenched and ended well before that of their US counterparts, and
TotalEnergies played a relatively more limited role in it and, for instance
left trade groups—including the American Petroleum Institute—over
climate policy. On the other hand—and perhaps most importantly—
Europeans, possibly for cultural reasons, seem not yet willing, or indeed
less willing than counterparts on the other side of the Atlantic, to employ
legally binding initiatives. Therefore, the implementation of the duty
of reparation needs a deeper engagement from the agents of destabilization. However, since the awareness of and sensitivity to the role
and responsibility of the fossil industry in climate change in the countries where major European IOCs are headquartered—especially in the
United Kingdom and the Netherlands—is already heightened and, by
and large, similar to that in the United States, primary agents of destabilization are not necessarily required to further their efforts greatly. It is
the operational agents—groups of citizens, nongovernmental organizations, social movements, and political authorities at various level—able

to bring oil and gas companies to the courts of justice that should engage in further efforts. Initiatives based on consumer and investor fraud over climate risks would, in fact, be of utmost use in inducing governments to take binding legal action to implement the duty of reparation, as chapter 8 explains. To move the political context toward this objective, a few years—at the very least—are needed. Once the optimal state of affairs has been reached, BP and Shell should be compelled to fulfill the disgorgement benchmarks reported in table 9.3. TotalEnergies, given its relatively less morally compromised situation, should possibly be granted a little more time to comply with the duty of reparation.

The second aggregation involves HR NOCs: CNPC/PetroChina, Gazprom, and Saudi Aramco. These companies and their contexts are different, although probably less so than many may imagine with regard to the issues at stake here, and their implementation of the duty of reparation should take account of this diversity. In terms of the elements that influence Big Oil's realization of the duty of reparation underlined in chapter 7, these three companies—besides having some of the biggest assets and among the highest disgorgement amounts, as shown in table 9.3—share a societal context unfavorable to action against fossil fuels; a rather solid and effective (especially in China) institutional system; the authoritarian nature, albeit in different forms and to different degrees, of the states to which they belong (Economist Intelligence Unit 2021); and a relatively acceptable economic situation. They differ, though, in terms of resource availability: CNPC/PetroChina is a resource seeker, while Gazprom and Saudi Aramco are market seekers, but Gazprom mostly operates with a fossil fuel (gas) that is more difficult to handle than Saudi Aramco (oil). Additionally, CNPC/PetroChina and Saudi Aramco have a broad mandate, and their main goal is basically limited to profitability, whereas Gazprom has political and social functions, such as providing subsidized gas domestically, besides its commercial goals (Victor, Hults, and Thurber 2012a). Therefore, on the one hand, the potential of compliance to the duty of reparation is greater for the market-seeking Saudi and Russian companies, given that they have fossil fuel reserves (at least in

the short to medium term) to rely on to meet the disgorgement requirements, especially Saudi Aramco; their duty of reparation is not mitigated
by their fossil fuel–exporting country status, since Russia and Saudi Arabia are, respectively, upper-middle-income and high-income countries in
terms of GDP. On the other hand, a major stumbling block for these
companies is that they are owned by states where the societal context is
patently favorable to the oil complex; China seems, however, more likely
to address the climate crisis by targeting fossil fuels and their producers.

Altogether, it seems possible to argue that to encourage CNPC/Petro-
China to satisfy its duty of reparation, a larger involvement of operational
agents of destabilization would be required, although their activities
could be greatly hindered or even prevented by the Chinese authoritarian
regime if it were to clash with the general national energy and climate
strategies. On the other hand, Gazprom and Saudi Aramco's duty of reparation would require a massive social effort brought about by primary
agents of destabilization. It should be noted again, however, that the
potential clout of the agents of destabilization is inversely proportional
to a country's level of authoritarianism; therefore, their role for inducing
governments to adopt legal provisions for the duty of reparation is actually very limited in the countries considered.

In sum, while all three companies considered—CNPC/PetroChina,
Gazprom, and Saudi Aramco—should disgorge a figure close to their
ideal benchmark shown in table 9.3, CNPC/PetroChina's implementation of the duty of reparation could be rolled out in a shorter time span,
provided there is sufficient consistency in anti–fossil fuel efforts with Chinese national strategies and plans. Gazprom and Saudi Aramco's compliance with their duty of reparation is still unlikely, at least in the short
to medium term. If, on the one hand, Russia's societal context allows for
a possible future destabilization of the oil complex, Saudi Arabia's, which
is basically built around oil, does not. It is not possible here to enter into
these countries' energy and fossil fuel policies or, more broadly, into their
industrial and economic ones, but unless radical transformations of the
global fossil fuel regime take place, they seem unwilling to rethink their

fossil fuel–based economies and move toward green models of economic growth in the near future. Therefore, it is difficult to imagine their fairly rapid achievement of the duty of reparation.

The third group of top twenty oil and gas companies includes a rather heterogeneous pool of NOCs: Abu Dhabi National Oil, Lukoil, Kuwait Petroleum, National Iranian Oil, Nigerian National Petroleum Corp, PDVSA, Pemex, Petrobras, Petronas, Rosneft, and Sonatrach. First and foremost, these companies should be granted as much leeway as possible in the disgorgement of their benchmarks (see table 9.3), especially those based in more impoverished countries—National Iranian Oil, Nigerian National Petroleum Corp, and Sonatrach—in view of the respect of a country's right to development. The diversity of these companies in terms of determinants that influence the implementation of the duty of reparation impedes the drafting of similar political avenues to fulfill it; furthermore, the lack of homogeneity makes it impossible to envisage a common role for agents of destabilization in triggering the necessary legal initiatives. At the same time, it is largely outside the scope of this book—and probably altogether impossible—to provide a detailed analysis of the possible evolution of the political and socioeconomic contexts of the countries and regions involved with regard to fossil fuels, climate change, and the part that oil companies have in it. Suffice it to say that a prominent role rests on the shoulders of primary agents of destabilization, despite the many difficulties they would face in the authoritarian regimes that most of the companies considered belong to.

These companies are based in five very different regions: Africa, the Arabian Gulf, Latin America, Russia, and Southeast Asia. With regard to the potential of agents of destabilization in Russia and the Arabian Gulf, the considerations put forward above are valid also for this third aggregation of NOCs: in a nutshell, the societal context is highly unfavorable to initiatives supporting and implementing the duty of reparation. Given the authoritarian nature of Algeria and Nigeria (Economist Intelligence Unit 2021), similar considerations also apply to Sonatrach and Nigerian National Petroleum Corp. In Latin America and South East Asia, however, primary agents of destabilization could indeed play a role, one that could prove to be more effective: Brazil, Mexico, and Malaysia have civil societies that show more sensitivity to the harmful effects of fossil fuels (but Pemex, since the end of 2020, is no longer active in the Oil and Gas Climate Initiative, the oil industry's key climate group). At the same time, the Development Bank of Latin America and the Asian Development Bank have sent out clear messages that they aim to stop funding fossil fuel projects. In brief, these regions could prove to be the next frontier of climate activism, so their NOCs could, in the medium term, be confronted with various demands for them to shoulder financial burdens in line with those required by the duty of reparation. Given the profound social, political, and economic crisis that Venezuela has long experienced, nothing meaningful about PDVSA in relation to the duty of reparation can be envisaged.

More interesting is to briefly highlight some of the possible repercussions that a duty of reparation for Pemex, Petrobras, and Petronas would have on societies based on their current nonproductive roles, ones stretching beyond fossil fuel production assigned to them by their governments. According to Victor, Hults, and Thurber (2012b, figure 20.3c, 899), NOCs may have different levels of burdens in terms of the provision of *social and public goods* (burdens that benefit society at large) and *private goods* (burdens that benefit narrow groups). Mexico's Pemex has a high burden in terms of social and public goods (high taxes whose revenues are employed by governments for broad public purposes) and an upper-middle burden in terms of private goods (patronage to labor unions), while Brazil's Petrobras has a lower-middle burden in terms of public and social goods (tools for energy self-sufficiency and to supply domestic markets) and a low private goods burden. Therefore, the Brazilian NOC, being much less constrained and financially limited than Pemex, should in principle be more forcefully compelled to meet its duty of reparation. Similarly to Petrobras, Malaysia's Petronas—on the lower end of the scale of burdens—should be expected to more swiftly comply with a duty of reparation once society lays the groundwork for the enactment of legal provisions.

Financial guru Jim Cramer's dire description of the failing allure of Big Oil stocks reported in chapter 8 is no fly-by-night comment. Three telling pieces of evidence suggest that it is not only oil itself that risks becoming a fossilized relic from a bygone era. Money-spinning superstars—oil and gas companies—may also pretty soon become antiquated curios if they do not shed their skin. The energy sector weighting in the S&P 500 hit a low of 3.8 percent at the end of March 2020; after ninety-two years, in late August 2020 ExxonMobil left the exclusive Dow Jones Industrial Average index, while equity issuances by fossil fuel producers in the period 2012–2020 have lost $123 billion in value and underperformed on the MSCI All Country World Index (ACWI) by 52 percent (CTI 2021c); and in January 2021 Tesla was worth more—that is, its stock market value, even if part of it might have been a bubble, was greater—than BP, Chevron, ConocoPhillips, ExxonMobil, Shell, and TotalEnergies put together.

A world that once revolved around oil is starting to falter. For instance, the spectacular ongoing technological revolution is changing its life-blood, energy, which was once almost exclusively produced with fossil fuels, and now renewables are playing catch-up at an unprecedented and unexpected rate: the total installed wind and solar capacity will surpass natural gas in 2023 and coal in 2024 (IEA 2020c).

The low-carbon transition—the overall lowering of carbon emissions from socioeconomic systems—is possibly one of the most powerful narratives underpinning current thinking about societies, economies, and nature in the face of the impending climate crisis, one that requires a rapid and profound modification of attitudes, behavior, norms, incentives, and politics.

This scientific, political, and socioeconomic debate, by and large, addresses the shift from a system dominated by finite yet easily available and relatively inexpensive fossil energy to one that progressively abandons fossil sources and moves toward renewables. As pointed out in chapter 8, one approach of social sciences useful in this context is that of transition studies, which explain how different strategies and resources influence the acceptance of social forces and the departure from current state of affairs. This chapter addresses the low-carbon transition of the specific socioeconomic force under scrutiny—that is, the oil industry—from this viewpoint. The duty of decarbonization imposed on the oil industry for its moral responsibility for climate change requires that it progressively reduces and eventually eliminate fossil fuels from its products and processes, thereby modifying its hue from the old black gold to a new green. Therefore, while limiting the carbon content of processes is inscribable in the broader goals of any sector in the context of the low-carbon transition, the decarbonization of this specific industry's products—oil and gas—has a unique and distinctive feature: it is the crucial variable, the first domino in the chain, for decarbonizing the entire global socioeconomic system, and this chapter is thus specifically devoted to the investigation of this point.

In light of these considerations, this chapter does not analyze or attempt to summarize the low-carbon transition; its more unassuming aim is to investigate the conceivable development of a part of it focused on specific agents, that is, how the progressive abandonment of fossil fuels demanded of Big Oil by the duty of decarbonization can be operationalized and implemented. As emphasized in chapter 9, this analysis is carried out in a realistic milieu of political evolution. Based on the useful four ideal-typical approaches to just transitions—status quo, managerial, structural, and transformative—Big Oil's duty of decarbonization requirements would belong to the structural reform approaches to just transitions, which entails institutional change and structural evolution through modified governance structures and broader participation (Köhler et al. 2019). As recalled in chapter 8, the unit of this analysis is at the mesolevel of socio-technical systems (Geels, Berkhout, and Vuuren 2016), which complements the macrolevel (e.g., changing the nature of capitalism or nature-society interactions, the *transformative just transition*) and the microlevel (e.g., changing individual choices, attitudes and motivations, which basically belong to the *status quo and managerial transition* approaches).

Additionally, it is worth clarifying that a broad analysis of the geopolitics of Big Oil's decarbonization is beyond the scope of this book, although relevant national and international political considerations are taken into account, especially with regard to national oil companies (NOCs). In fact, in petrostates, if decarbonization processes prevent their NOCs from generating the revenues needed to sustain the socio-economic system, they can be disruptive and produce political instability, which triggers different geopolitical scenarios (Goldthau et al. 2019). A low-carbon transition could, for instance, decrease by 51 percent petrostates' oil and gas revenues over the next two decades (CTI 2021a) and endanger the stability of those that are not bracing for it (Verisk Maplecroft 2021). Additionally, if such a transition occurs too rapidly, migration from oil-dependent regions to Western countries can increase, thus burdening what is already a hot topic in international and national politics. Riots, further extremism, and internal conflicts could be triggered, causing the basic structure of a state to fall apart with potentially dangerous regional and global consequences. Indeed, the geopolitical implications of the low-carbon transition will be very subtle, complex, and counterintuitive: petrostates, for instance, could temporarily profit from it, since, as demand peaks, it is the lowest-cost producers—such as the Persian Gulf NOCs—that will be able to sell their oil the longest (Bordoff 2020).

Focusing exclusively on the decarbonization of Big Oil is no mean feat, given the overall complexity and implications that it would have on national, regional, and global socioeconomic systems. For instance, the oil industry must substantially cut combined production to keep emissions within international climate targets while fighting for survival in an ideally more carbon-conscious world. To throw a few statistics in, the seven largest investor-owned oil companies (BP, Chevron, ConocoPhillips, ENI, ExxonMobil, Shell, and TotalEnergies) must reduce their production by 40 percent by 2040 if they want to stay below the International Energy Agency's (IEA) *"beyond 2 degrees" scenario* (IEA 2019b).

At any rate, downsizing an industry with $16 trillion worth of capital and at least ten million employees, which has already consumed up to 2019 roughly 82 percent of the 2,810 $GtCO_2$ total carbon budget for a 50 percent chance of success of staying below 1.5°C of global warming (CTI 2019b), requires a *herculean effort*. An even greater effort may be required, since many countries rely on substantial oil rents to finance public services (World Bank 2019), twenty-three countries get more than 50 percent of their export income from fossil fuels (Ross 2019), and some fossil fuels should nonetheless be supplied in the future, as certain products—mainly petrochemicals (e.g., plastics for medical use)—and industrial processes have yet to or cannot be decarbonized.

At the same time, while the IEA reports that in 2019 oil and gas industry fossil fuel capital investments in energy were 99.2 percent of the total compared to a mere 0.8 percent of those in renewables and carbon capture and storage (IEA 2020b), evidence confirms that the world's fifty biggest oil companies plan to flood the planet with an additional seven million barrels of crude oil per day over 2020s, since they have projected an increase in their production of more than 35 percent between 2018 and 2030, a much sharper intensification than in the past (Watts, Ambrose, and Vaughan 2019). This will be coupled by the envisaged inordinate amount of shale oil and gas, in line with the decade-long bonanza that, despite the spectacular unprofitability of these extraction techniques, saw the opening of 245,000 wells in the United States alone between 2009 and 2019 (Kelly 2020). New oil production is furthermore expected from Brazil, Canada, Guyana, and Norway (the Scandinavian country awarded sixty-one offshore exploration rights to thirty oil companies in January 2021), which are projected to add one million barrels per day to the currently produced eighty million starting from 2020 (Krauss 2019), while Russia and Suriname led new oil and gas discoveries in 2020. Such investments privilege fossil fuel reserves that can be productive in a short span of time rather than developing expensive far-flung reserves, given the expected long-term downward trend of fossil fuel prices and the possibility that they eventually become stranded (Jaffe 2020).

Additionally, the industry seeks to ensure a carbon-intensive future by expanding production of plastics (Corkery 2019): the IEA calculates that petrochemicals will account for almost half of the growth in oil demand up to 2050 (IEA 2018). But despite oil and gas companies betting on plastics, its demand may soon peak too as the economy begins to move from a linear to a circular plastic system (CTI 2020b).

In the face of the enormity and complexity of the challenge of decarbonizing the oil industry's processes and products, a colossal enterprise fraught by contrasting powerful interests and political and economic struggles, a managed decline of the industry is paramount. This chapter, building on the analysis conducted so far in the book, will attempt to frame the task within the requirements of the duty of decarbonization by first exploring the so-called lock-in dynamics in current fossil fuel–intensive behaviors and patterns as well as the agents of destabilization and instruments for escaping such carbon lock-ins. The chapter will then go on to explore the operationalization and implementation of the duty of decarbonization for the top twenty oil and gas companies. Finally, the closing section puts forward a pathway for orchestrating the industry's decarbonization efforts to become Big Green.

BIG OIL AND CARBON LOCK-INS

Based on the considerations carried out in the previous chapters of part III, it is quite straightforward to assume that a low-carbon transition is mostly a political and moral issue rather than simply a technological or institutional one whereby hegemony, power dynamics, distribution of and access to resources, and more generally matters of political economy as well as moral considerations about vulnerable people, groups, and communities are critical (Patterson et al. 2018). Political authorities, especially governments, have not so far cleared the path for the low-carbon transition; quite the contrary, they continue to back the fossil fuel industry through, for example, subsidies and support for oil and gas infrastructure (Roberts et al. 2018). At the same time, politics and justice suggest reorienting the low-carbon transition upstream, that is, to also—or indeed primarily—address producers of fossil fuels through supply-side measures instead of focusing on consumers through demand-side provisions (Lazarus and van Asselt 2018). It is, in fact, the oil complex that is the real nest of power, politics, and political economy and where a *just transition*—including impacts of fossil fuel production on humans and labor as well as on intergenerational and intragenerational justice issues—should be concentrated (Healy and Barry 2017).

Oil and gas companies are at the helm of the political power engine within the oil complex, wielding considerable influence on policy, as shown previously. So, what is required is a blueprint of political conditions to escape the different carbon lock-ins created and protected through such power and policy influence. For instance, the IEA in its 2020 *World Energy Outlook 2020* (IEA 2020e) points out that if energy infrastructures in operation and under construction were used in line with past practice until the end of their lifetimes, they would generate a level of emissions that would produce a long-term temperature increase of 1.65°C.

Ongoing investment in fossil fuel infrastructure and the inertia of institutions and of individual and social behavior bind society to carbon-intensive energy systems and patterns by creating assets, structures, and models that ensure future fossil fuel extraction and the inevitable associated emissions. This makes it harder for low-carbon energy alternatives and challengers to compete, creating a significant barrier to meeting climate protection goals. Addressing carbon lock-ins requires breaking the hold of the oil complex over political systems, institutions, and energy cultures. Additionally, when governments and other relevant agents address decarbonization through actions that create economic winners and prevent/weaken backlash from economic losers, it is more likely to destabilize the oil complex (Meckling 2019).

Measures that aim to tackle inequality and protect the weakest subjects are sorely needed (Bernstein and Hoffmann 2019; Aklin and Mildenberger 2020). They include, for instance, support to displaced workers, unemployment protection, placement support, and relocation grants; support to frontline communities whose main source of financial is fossil fuels; and protection to the weakest investors, such as pension funds. On the other hand, resistance to decarbonization strengthens lock-ins. Carbon lobbies and other entrenched powerful groups in the company's administrative or operational headquarters might actively hamper or slow down processes of decarbonization. The lower the internal resistance, the more likely the process of decarbonization will see the way ahead paved smoothly.

The relevant literature evinces three types of carbon lock-ins: technological/infrastructural, institutional, and behavioral (Seto et al. 2016). The technological/infrastructural lock-in is basically determined by the constraints imposed by prior decisions relating to carbon-based technologies, infrastructures, practices, and their support to future carbon-intensive paths, making it more challenging or even impossible to subsequently pursue more suitable paths toward low-carbon socioeconomic systems (Erickson et al. 2015a, 2015b).

Institutional lock-in refers to the circumstance that institutional choices made at any given point in time shape institutions' subsequent choices. Such lock-ins reflect the political conflict between agents who benefit from the existing set of economic, social, and cultural arrangements that favor carbon-intensive socioeconomic systems and those who would instead benefit from decarbonized ones (Seto et al. 2016). In fact, given the power of the oil industry and the close relationships between governments and the companies within the dominant oil complex, institutions tend to choose and act in ways favorable to the oil industry, thus deepening and strengthening the vicious circle of the institutional lock-in.

Behavioral lock-in depends on patterns of human behavior and is divided into the lock-in of carbon-intensive behavior through individual decision making based on individual cognitive processes (not relevant for the purposes of the current analysis) and the lock-in dependent on social structures and practices determined by routines and norms embedded in the wider sociotechnical environment.

Big Oil is so shielded and protected by these lock-ins that its decarbonization must first and foremost confront the political dynamics that have produced and continue to reinvigorate them. And there lies the rub: to dismantle carbon lock-ins, what is required are disruptive, game-changing social, economic, and political innovations with bottom-up, participative approaches and top-down, centralized technocratic measures: two irreconcilable approaches, it would seem. However, addressing Big Oil's lock-ins through agents of destabilization may kill two birds with one stone, as this approach would fulfill the demands for participation hailing from society, with the effectiveness of coordinated widespread actions put into practice by other operational agents of destabilization working at different levels with diverse approaches.

Indeed, technological/infrastructural, institutional, and behavioral carbon lock-ins are mutually interdependent: they occur through the combined interactions between technological/infrastructural systems, governing institutions, and conduct and activities associated with energy-related goods and services (Unruh 2000). For analytical purposes in relation to the operationalization of Big Oil's duty of decarbonization, however, it is worth maintaining these three types of lock-in as subdivisions. The following section clarifies the instruments that agents of destabilization can use to more effectively dismantle the lock-ins outlined above. It is thus possible to lay the groundwork for all options of routes leading to the low-carbon requirements demanded of Big Oil by the duty of decarbonization.

It is worth recalling that the low-carbon transition is already happening now, and in contrast to the duty of reparation, Big Oil could voluntarily take larger strides toward greener business models that would be consistent with the objective of the duty of decarbonization; in other words, Big Oil's industrial goals could themselves generate ruptures in its carbon lock-ins. The impact of this prospect should not be overemphasized, and should, in any case, be considered incognizant, endogenous shifts of the oil complex that can reinforce the cognizant, exogenous destabilizations analyzed in the ensuing section.

OPERATIONALIZING BIG OIL'S DUTY OF DECARBONIZATION: AGENTS OF DESTABILIZATION AND INSTRUMENTS FOR ESCAPING CARBON LOCK-INS

To address the duty of decarbonization, one starting point is unavoidable: understanding how to plan a suitable foray into Big Oil's fossil fuel fortress. More specifically, to implement the duty of decarbonization, its operationalization must first be framed. This can be achieved by investigating the potential of agents of destabilization as well as the instruments they can deploy to overcome the carbon lock-ins. Once an escape route, so to speak, out of lock-ins is defined, the model of the duty of decarbonization for the twenty top oil and gas companies—taking into account their different levels of requirements as well as the determinants indicated in chapter 7—can be implemented. Accordingly, this section attempts to invoke Big Oil's duty of decarbonization by investigating agents of destabilization and instruments that can effectively erode carbon lock-ins; the ensuing section looks at how it can be implemented.

Keeping in mind the analytical distinction between primary agents of destabilization, who prepare the terrain by creating new social/moral norms and undermining resistance, and operational agents, who aim at changing oil and gas companies' behavior through different instruments, it is the latter group that can prove to be of most value with regard to technological/infrastructural and institutional lock-ins. In these contexts, the effectiveness of operational agents' actions is, of course, complemented and amplified by primary agents' diffusion of knowledge and raising awareness about the inadmissibility of fossil fuels as well as ways to phase them out.

On the contrary, the behavioral lock-in—individual and social—is the domain of primary agents of destabilization: those with an influential and catalyzing effect on behaviors and mindsets, the so-called norm entrepreneurs, such as religious leaders, writers, screen actors and broadcasters, influencers, gifted communicators, scientists, and norm champions including environmental advocacy groups, reliable investigative media sources, and social movements. The instruments and strategies to be employed are explained in chapter 8, so it would be gratuitous to detail them again here: the norm-spreading processes and the undermining of resistance for a general destabilization of Big Oil are the same as those required for dismantling the behavioral lock-in.

Suffice it to say that actions to dismantle Big Oil's behavioral lock-in should preferably aim to erode the naturalization of the use of fossil fuels described in chapter 5, which has endured despite the unanimous scientific acknowledgment of their harmfulness. They should also attempt to undermine the prevailing reactionary rhetorical arguments used in defense of fossil fuels, usually based on the dire economic effects of their dismissal, and the disruptive potential of supply-side climate policy and of divestment as well as the importance of low-carbon consumerism and lifestyles. These norm-spreading actions prove to be useful for the other lock-ins too: on such fecund terrain, operational agents can more easily and effectively introduce actions targeting the technological/infrastructural and institutional lock-ins that protect Big Oil.

Breaking free from the carbon trap requires more than just switching to low-carbon technologies or building the necessary infrastructure; nevertheless, the technological and infrastructural lock-in is still the fundamental first step. Given the urgency of the climate crisis, this lock-in should be addressed through an approach of *discontinuity*, which seeks a rapid transition to a different technological/infrastructural system characterized by radical changes (Unruh 2002). In terms of technology, the potential and know-how already exist, as the recent reports of the IEA indisputably show (IEA 2020b, 2020c, 2020e, 2021c).

Much effort is required, though, as a climate-safe system calls for $110 trillion worth of investments in the energy sector by 2050; currently $95 trillion has been earmarked, but mostly for fossil fuel investments. This mammoth sum should be redirected to clean technologies. In this regard, therefore, the operational agents of destabilization seem to be mostly economic agents and, in particular, the financial sector: commercial banks, development banks, insurers, pension funds, and sovereign wealth funds. They should become further inclined to progressively abandon the funding of fossil fuel projects in favor of low/zero-carbon activities, as shown in chapter 8. This trend could significantly benefit from supply-side climate policies banning certain types of fossil fuel production (e.g., fracking) or the phasing out of fossil fuels altogether, as some political local authorities including the US states of Hawaii and California (Roth 2019) and countries as Costa Rica and Denmark, which are building a coalition—the Beyond Oil and Gas Alliance—to bring an end to oil and gas production (Leylim 2021), have already done or planned to do.

Research institutions are another important operational agent of destabilization in relation to the technological lock-in. They can develop new products, services, and business models and create the markets for such technologies as well as diffuse them. On a different level, research institutions should also try to operate as primary agents of destabilization to steer and shape societal discourse, problem framing, and collective expectations on new low/zero-carbon technologies. It is worth recalling that the success of operational agents of destabilization in this context relies greatly on the vital and effective work—as clarified in chapter 8—carried out by all primary agents of destabilization in promoting the relevance of fossil fuel divestment as well as for prompting and organizing social/justice movements forcefully demanding it.

Financial agents of destabilization also play a prominent role in the infrastructural aspect of this kind of lock-in. This is a thorny issue, though for instance, the petroleum products and natural gas pipelines industry will see $88.4 billion and $78.8 billion in investments pumped into their industries, respectively, in the United States alone (GlobalData 2018). Given this ominous trend, a ban on the development of new fossil fuel infrastructures—a main supply-side measure—imposed by political authorities would greatly benefit the transition.

Escaping the suffocating embrace of fossil fuels and carbon entanglement in the broadest sense requires concentrating on the institutional lock-in, especially on the power relations between political authorities and the oil industry. Institutions here are understood as being distributional instruments oriented and constrained by considerations of power. It is easy, then, to see why, in the optic of this book, the institutional lock-in is the most entrenched and difficult of carbon lock-ins involving Big Oil. The institutional lock-in is the fruit of the conscious efforts deployed by the hegemonic oil complex to create and maintain its power and influence, as clarified in chapter 7. The companies that make up the complex have intentionally built a resilient regime, which protects and perpetuates the carbon-intensive status quo through intentionally coordinated efforts to structure policies, rules, norms, instruments, and constraints thus, their goals and interests are safeguarded. In this context, however it is important to recall the all-important role of primary agents of destabilization to help overcome institutional lock-in with initiatives able to "galvanize stakeholder attention" (Seto et al. 2016, 435), such as those carried out by the mentioned charismatic norm entrepreneurs (e.g., Pope Francis and Greta Thunberg) and champions (e.g., divestment and climate justice movements).

In its essence, institutional lock-in largely reflects the struggle between a Big Oil that overwhelmingly benefits from the carbon-intensive status quo versus agents benefiting from a decarbonized socioeconomic system. Escaping such lock-in requires making agents who benefit from decarbonization economic winners, safeguarding their benefits against a backlash from Big Oil, which could, if not actually become an economic loser, to some extent slide down the winner/loser scale.

To this end, the main operational agents of destabilization are political authorities at various levels, through a portfolio of different targeted demands and mostly restrictive supply-side instruments and policies, including carbon pricing, subsidy reduction, production quotas, supply ban/moratorium, support for clean energy research and development, supply taxes, subsidies, tax rebates, loan guarantees, and deployment mandates for renewables.

To nurture those who could benefit from a decarbonized world and to safeguard them from the backlash of the incumbent oil industry, political authorities face two main challenges. First, they need to build long-term political support for low-carbon initiatives by enacting policies that expand economic opportunities to other sectors too, thus creating a mutually supportive coalition of businesses, workers, and individuals able to disrupt fossil fuel dependence (Bernstein and Hoffmann 2019). Opportunities of decarbonization should be created in order to mobilize well-organized and powerful interests and to generate more solid feedback dynamics around a low-carbon energy system. Second, political authorities must address the direct and indirect costs that decarbonization creates to the oil industry to avoid or limit retrogressive actions; this should take place at each level of governance and across multiple economic and social sectors. Oil and gas companies would, for instance, bear the brunt of profit losses, as they would lose a share of the market. Therefore, political authorities should focus on both cost containment and ways of counteracting or weakening any opposition the oil industry may make (Meckling 2015; Meckling et al. 2019).

To further reinforce the decarbonization, political authorities should also foster a positive institutional lock-in in a new decarbonizing trajectory

(Seto et al. 2016). One possibility is increasing competition in the energy sector through feed-in tariffs so that the fossil fuel companies more aligned with serious decarbonization objectives can participate in a low-carbon future scenario.

IMPLEMENTING THE DUTY OF DECARBONIZATION: BIG OIL'S CONTRIBUTION TO THE LOW-CARBON TRANSITION

An important study (Rogelj et al. 2015) shows that to keep warming below 2°C by the end of the century, current global emissions need to be halved by the late 2030s and reach zero by around 2065. The 1.5°C objective requires that emissions be reduced by half by the early 2030s and reach zero by 2050. These estimates rely on negative emissions tech nology that is as yet unproven and unavailable at scale; otherwise, the trajectories of abatements must be substantially anticipated.

At any rate, to achieve such goals, fossil fuel use and, consequently its extraction and production must decline at more or less the same rate (Muttitt et al. 2016). For instance, according to a study by the Carbon Tracker Initiative (CTI 2019a), which sets fossil full companies' limits consistent with the Paris Agreement goals, seven among the major international oil companies (IOCs)—BP, Chevron, ConocoPhillips, ENI ExxonMobil, Shell, and TotalEnergies—must cut their emissions by 40 per cent and production by 35 percent by 2040 to stay within their company-level carbon budgets based on the IEA's *"beyond 2°C" scenario* for a rapid decarbonization pathway in line with international policy.

The scientific, policy, and business communities have not been slow in drafting countless hypothetical decarbonization scenarios to achieve these objectives. The IEA in its *World Energy Outlook 2019* report (IEA 2019c) uses the *Sustainable Development Scenario* (SDS), which "holds the temperature rise to below 1.8°C with a 66% probability without reliance on global net-negative CO_2 emissions; this is equivalent to limiting the temperature rise to 1.65°C with a 50% probability. Global CO_2 emissions fall from 33 billion tonnes in 2018 to less than 10 billion tonnes by 2050 and are on track to net-zero emissions by 2070" (IEA 2020d).

The SDS can be taken as an ideal benchmark for Big Oil to meet its duty of decarbonization for at least four reasons. First, it sets out a sufficiently ambitious and yet pragmatic vision of the achievement of critical sustainable development goals, consistent with the realistic reference setting of the book.

Second, the IEA belongs to the global oil establishment; indeed, it is sometimes accused of being too conservative. Therefore, its scenarios are never meant to over-penalize the oil industry. The research, communications, and advocacy organization Oil Change International claimed that "the IEA has again failed where it matters on climate. . . . Without stepping up and making high ambition the centrepiece, the IEA seems to be confirming they are not fit for purpose in a time of climate emergency. . . . The IEA should be guiding the world away from the climate crisis. Unfortunately, the IEA has failed to convey the urgency of the situation" (OCI 2019).

Third, the IEA SDS does not rely on net negative emissions (unlike, for instance, the IPCC 1.5°C scenarios, eighty-eight out of ninety of which assume some level of net negative emissions). The current analysis, in fact, is not focused on a fully fledged investigation of societal decarbonization; therefore, the inclusion of approaches based only on abatement measures makes it possible to apply the indications of this scenario to a specific source of emissions such as oil and gas companies.

Fourth, the alleged lack of stringency of this scenario implies that it does not impose lethal emission cuts to the oil industry and thus guarantees its immediate economic survival while preserving its capacity to meet its duty of reparation. In other words, the use of the IEA SDS can in this context of analysis be considered superior to more stringent scenarios, as it implicitly takes into account the trade-offs between the duties of reparation and decarbonization, and while sufficiently driving the oil industry toward the latter, it does not hinder it from achieving the former, as argued in chapter 6.

Basically, consistent with the SDS, Big Oil is required to reduce its scope 1 and 3 emissions by roughly 70 percent by 2050; the SDS envisages global CO_2 emissions falling from 33 billion tonnes to less than 10 billion tonnes by 2050. In this view, there is a lenient assumption that 30 percent of Big Oil's current emissions are allowed for irreplaceable uses, so to speak, while allowing the industry to secure the necessary financial means to satisfy the duty of reparation. This *mild* assumption, added to the choice of not adopting more ambitious decarbonization scenarios, makes the duty of decarbonization less onerous and more feasible in the long term.

As a further simplification, given the exemplary goal of the current exercise, the timeline for decarbonization is not considered, but only the final goal of a 70 percent abatement by 2050, achievable through linear cuts consistent with the objective defined by the IEA SDS, is considered. Indeed, the ways and means required to meet this objective would be left to the discretion of each individual company. Finally, a discrepancy in the data presented should be pointed out. Data on oil and gas companies' scope 1 and 3 emissions are available only up to 2015, whereas the IEA SDS starts from 2018; however, given the analytical goal of the figures indicated in this chapter, this does not undermine their indicative potential. Table 10.1 provides the 2015 scope 1 and 3 emissions and the 2050 scope 1 and 3 *target* emissions for the top twenty oil and gas companies.

To balance the somewhat alleged lack of meaningful ambition of the IEA SDS, oil and gas companies are further required to comply with a *managed decline scenario* whereby "no further extraction infrastructure is developed, existing fields and mines are depleted over time, and declining fossil fuel supplies are replaced with clean alternatives" (Muttitt et al. 2016, 32). This is consistent with the findings of a landmark report of the International Energy Agency (IEA 2021c), which suggests that to achieve the 1.5°C target a rapid decrease of fossil fuel production as a result of no new project is necessary.

This is a further requirement that, by barring new fossil fuel projects and managing the decline of the oil industry over time, would avoid the worst impacts of climate change and increase the chances of Big Oil's

Table 10.1

Top 20 oil and gas companies' scope 1 + 3 greenhouse gas emissions 2015 and *target emissions* in 2050 (2050: 70% reduction compared to 2015), $MtCO_2e$

Oil and Gas Company	2015	2050
Saudi Aramco	1,951	585
Gazprom (Russia)	1,138	341
National Iranian Oil	1,036	311
Rosneft (Russia)	777	233
CNPC/PetroChina	625	188
Abu Dhabi National Oil–ADNOC	584	175
ExxonMobil (USA)	577	173
Pemex (Mexico)	530	159
Shell (UK/Netherlands)	508	152
Sonatrach (Algeria)	492	148
Kuwait Petroleum	478	143
BP (UK)	448	134
PDVSA (Venezuela)	398	119
Petrobras (Brazil)	382	115
Chevron (USA)	377	113
Petronas (Malaysia)	340	102
Nigerian National Petroleum Corp	329	99
Lukoil (Russia)	328	98
TotalEnergies (France)	311	93
ConocoPhillips (USA)	224	67
Top 20 O&G Companies	**11,833**	**3,550**

Source: Author's elaboration from the Carbon Majors Database—2017 Dataset Release (CDP 2017). According to the Greenhouse Gas Protocol of the World Resources Institute (WRI n.d.), scope 1 emissions refer to direct oil and gas combustions; scope 3 emissions originate from the downstream combustion (for energy and nonenergy purposes) of oil and gas that they have distributed within the global economic system (the largest share, roughly 90%, of oil and gas companies' emissions are scope 3).

compliance with the goals of the Paris Agreement (Scott 2018). Big Oil is therefore expected to abandon all future fossil fuel projects.

Besides stopping any future investment in fossil fuels, ambitious abatement objectives are required that involve cutting scope 1 and 3 emissions associated, respectively, to their processes and products by 70 percent in 2050. According to Heede (2013, 2014), scope 3 emissions amount roughly to 90 percent of the emissions associated to the industry, with scope 1 emissions accounting for the remaining 10 percent; table 10.2 reports scope 3 target emissions for the top twenty oil and gas companies. For the sake of the analysis conducted here, the reduction in scope 3 emissions should be understood as the very objective of the duty of decarbonization as it testifies to a willingness to abandon fossil fuels, as required by a managed decline scenario. Scope 1 emissions abatement, despite its importance in the overall picture, resonates more with a business-as-usual low-carbon scenario, which does not involve a structural change in the business model of the company. At any rate, the oil industry actually has several effective and efficient options available to address scope 1 emissions, which in part it seems willing to adopt (IEA 2020a).

Therefore, the figures reported in table 10.2 represent the "absolute target values" of the effort required by the duty of decarbonization in terms of abandonment of carbon-intensive *products* by the top twenty oil and gas companies. Absolute target values mean here the emissions associated with the fossil fuels sold to the global economy; from a different perspective, as these emissions correspond to a specific quantity of fossil fuels, the values reported in table 10.2 are proportional to the quantity of fossil fuel products that oil and gas companies must stop producing in order to meet the requirements of the IEA SDS.

As in the case of the indicative disgorgements required by the duty of reparation suggested in the previous chapter, the 2050 scope 3 target emissions provided in table 10.2 should be seen as an ideal to which all companies must aspire, despite not all of them necessarily being required to fully comply with it. In fact, the fulfillment of these reductions is influenced by the social, political, and economic factors of the determinants that influence Big Oil's duties, as outlined in chapter 7: societal context, institutional strength, economic and political situation, and resource availability and nature of the resource. To better understand the top twenty oil and gas companies' level of commitments demanded by the duty of decarbonization, it is vital to attribute greater weight to the

Table 10.2

Top 20 oil and gas companies' scope 3 greenhouse gas
target emissions in 2050, $MtCO_2e$

Oil and Gas Company	2050
Saudi Aramco	527
Gazprom (Russia)	307
National Iranian Oil	280
Rosneft (Russia)	210
CNPC/PetroChina	169
Abu Dhabi National Oil–ADNOC	158
ExxonMobil (USA)	156
Pemex (Mexico)	143
Shell (UK/Netherlands)	137
Sonatrach (Algeria)	133
Kuwait Petroleum	129
BP (UK)	121
PDVSA (Venezuela)	107
Petrobras (Brazil)	103
Chevron (USA)	102
Petronas (Malaysia)	92
Nigerian National Petroleum Corp	89
Lukoil (Russia)	89
TotalEnergies (France)	84
ConocoPhillips (USA)	60
Top 20 O&G Companies	**3,195**

Source: Author's elaboration from the Carbon Majors Database—2017
Dataset Release (CDP 2017).

objective parameter of *contribution* employed to group the various companies in chapter 7. Such determinants and the relevance of the contribution parameter testify to the capacities of the sociopolitical and economic context the oil company finds itself in as well as of the company itself in creating or already being in possession of the conditions of breaking free of carbon lock-ins and thereby transitioning to low-carbon business models.

IOCs

Despite grouping oil and gas companies for the operationalization and implementation of the duty of reparation, as shown in chapter 7, the duty of decarbonization needs to take account of the distinction between IOCs and NOCs for one very straightforward reason: while NOCs' revenues contribute to social and public goods as well as to the provision of private goods in their home countries, IOCs—besides having quite similar situations with regard to the determinants mentioned above—respond only to their shareholders in economic terms. In other words, as they do not have mandatory social functions and do not face the associated economic constraints, IOCs should fully meet the benchmark of decarbonization associated to their scope 3 target emissions set out in table 10.2. The levels of abatement required would allow them the possibility to simultaneously fulfill their duty of reparation, as stressed above. For the sake of clarity, IOCs' requirements in terms of actual emissions levels by 2050 as demanded by their duty of decarbonization are grouped together in table 10.3, excluding NOCs. It should be noted that while ConocoPhillips and TotalEnergies belong to the MR grouping, as table 7.2 shows, their private ownership and consequent lack of social functions—i.e., the fact that they are IOCs—mean that these companies nonetheless are among those with the most stringent duty of decarbonization.

NOCs

In the case of NOCs, the implementation of the duty of decarbonization needs a more exhaustive analysis that can be usefully carried out within the HR/MR/LR groupings of chapter 7. It was stressed there that given their social functions, NOCs' duty of decarbonization should be more prudent than IOCs' and that emphasis should be on the contribution parameter goal.

Based on these assumptions, it is possible to argue that with specific regard to NOCs, ADNOC, CNPC/PetroChina, Gazprom, Kuwait Petroleum, and Saudi Aramco belong to the HR group; Lukoil, Pemex, Petrobras, Petronas, and Rosneft to the MR group; and National Iranian Oil, Nigerian National Petroleum, PDVSA, and Sonatrach to the LR group (see

Table 10.3

IOCs' scope 3 greenhouse gas emissions levels in 2015 and 2050 (30% of 2015) and emissions to be abated in the period 2015–2050, $MtCO_2e$

IOCs	2015	2050	Abatements
ExxonMobil (USA)	519	156	−364
Shell (UK/Netherlands)	457	137	−320
BP (UK)	403	121	−282
Chevron (USA)	339	102	−238
TotalEnergies (France)	280	84	−196
ConocoPhillips (USA)	202	60	−141
Total	2,201	660	−1,540

Source: Author's elaboration from the Carbon Majors Database—2017 Dataset Release (CDP 2017).

Table 10.4

NOCs grouping in relation to the duty of decarbonization

High Requirement (HR)	Medium Requirement (MR)	Low Requirement (LR)
ADNOC, CNPC/ PetroChina, Kuwait Petroleum, Gazprom, Saudi Aramco	Lukoil, Pemex, Petrobras, Petronas, Rosneft	National Iranian Oil, Nigerian National Petroleum, PDVSA, Sonatrach

Source: Author's elaboration from the Carbon Majors Database—2017 Dataset Release (CDP 2017).

table 10.4). It should be noted that in this case the general grouping indication of chapter 7 is not respected because among determinants, besides attributing greater weight to the contribution parameter goal, it is believed that the economic one is of cardinal importance.[1] Additionally, no attempt to quantify the *generousness* in terms of a *discount* on the 70 percent decrease in 2050 emissions of IOCs is carried out, as it would inevitably be fraught with overly subjective considerations: abatement commitments are rather described in qualitative terms. Suffice it to say that in general, the HR, MR, and LR groups are granted progressive reductions on the ideal 70 percent objective or lengthier time horizons than 2050 (no longer than 2070, though, the IEA SDS net-zero emissions target year).

In relation to HR NOCs, with regard to the determinants of chapter 7, Russia and Saudi Arabia share a similar societal context adverse to decarbonization; China, Kuwait, and particularly Abu Dhabi seem increasingly interested in finding alternatives to fossil fuels. In economic terms, the Persian Gulf states are solid, whereas the others have more limited wealth; politically, all countries with NOCs in the top twenty can be considered authoritarian (Economist Intelligence Unit 2021) with a sufficient level of institutional stability; China and the Gulf states have the potential to deploy renewables at scale, despite currently having a high dependence on fossil fuels.

This necessarily cursory overview suggests that ADNOC and Kuwait Petroleum should bring their decarbonization as close as possible to the levels indicated in table 10.2. CNPC/PetroChina, Gazprom, and Saudi Aramco should instead have a reduced obligation, in terms of time and scale, to decarbonize their products. From a different perspective, the emergence of a favorable political and social context to promote and implement legal initiatives for decarbonizations can be foreseen in Abu Dhabi and Kuwait. This is another factor that explains the bar being higher in terms of abatements for ADNOC and Kuwait Petroleum.

As for MR NOCs, the determinants considered show a certain degree of internal consistency across all countries. For instance, their GDP per capita varies from $10,192 for Malaysia to $9,972 for Russia, $8,069 for Mexico, $6,450 for Brazil (all data refers to the International Monetary Fund 2020 GDP per capita values at current prices [IMF 2021], as specified for the HR NOCs above). All in all, it seems that their efforts to decarbonize should be slightly inferior to those of the third subgroup of HR NOCs above, that is, Gazprom and Saudi Aramco.

Similarly, LR NOC countries should be considered as a rather homogenous group with regard to the determinants shaping the duty of decarbonization. Therefore, National Iranian Oil, Nigerian National Petroleum, PDVSA, and Sonatrach should be granted the largest *discount* compared to their 2050 target values and the longest possible period for carrying out the envisioned abatements.

A PATHWAY TO BIG GREEN

Fossil fuel advocates loudly proclaim how the low-carbon transition entails massive costs, especially in terms of job losses, increased/regressive energy prices, dangers to economic growth, and loss of revenues; on the other hand, campaigners for a low-carbon future underline the costs associated with the incumbent fossil fuel regime: the impending climate crisis, pollution, exploitation, corruption, conflicts, and violence. A balanced vision would be to admit that a low-carbon transition generates benefits and costs, broadly understood, borne by different agents: it creates *winners* and *losers* and therefore involves significant moral issues (Newell and Mulvaney 2013, 133).

The duty of decarbonization—Big Oil's main contribution to the low-carbon transition—similarly involves moral issues. In this perspective, the aim of this final section is first to underline the general moral principles that should underpin oil and gas companies' duty of decarbonization. Justice, by providing unifying moral principles, plays a major role in facilitating collective action in issues of this kind. The more the duty of decarbonization is informed by moral principles, the more a managed decline of the oil industry's involvement in fossil fuels can, in principle, be achieved. Furthermore, based on such principles, this section suggests a possible pathway for Big Oil to transition into Big Green.

The idea of a *just transition* was first developed by trade union and environmental/climate justice movements and was mainly focused on the job losses that would be caused by a reduction of fossil fuel production. The objective was to provide displaced workers and frontline communities with appropriate job opportunities and other context-specific forms of assistance as well as to create the conditions for their participation in the entire transition process (UNFCCC 2016).

Big Oil's duty of decarbonization requirements, however, entails a broader understanding of just transition, which indeed includes considerations about job losses, given their complexity, sensitivity, and implications. In such terms, a just transition is a fair and equitable process of moving toward a low-carbon society. The different scholarships (e.g., climate, energy, transition research, environmental studies, social sciences) working on the low-carbon transition have diverse notions of justice and therefore different understandings of what exactly a just transition should involve. For instance, Heffron and McCauley (2018) argue that it should involve a comprehensive approach to three dimensions of justice: distributional, procedural, and restorative (this book refers to the latter approach as *corrective*, as clarified in part II). While corrective justice issues provide the general moral background for developing both Big Oil's duties, distributive and procedural justice are important to justly achieve the duty of decarbonization within a context of managed decline.

Specifically, the moral principles put forward in this section are essentially distributional; however, their realization must include considerations of procedural justice, that is, of the fair involvement of all interested parties in the schemes of collaborative social decision making required and produced by the duty of decarbonization. To this end, recognition and participation are key features of procedural justice. Suffice it to say here that procedural justice increases the practicability of the moral case requiring Big Oil to decarbonize its processes and products. At any rate, it is worth recalling that adequately addressing the moral concerns raised by Big Oil's duty of decarbonization is fundamental to fostering its feasibility.

There are two main moral concerns raised: to minimize the disruption of key developmental priorities, with regard to the provision of energy services and the possibility of diversification for the relevant economies, and to distribute its costs fairly. An additional moral concern is, necessarily, the one stemming from the original just transition demands focused on the protection of workers and communities (Kartha et al. 2018).

To address these moral concerns underlying the just achievement of Big Oil's duty of decarbonization, two overarching moral principles are required: proportionality and sufficiency.

Proportionality

In its broadest and least controversial understanding, the principle of proportionality demands a reasonable balance between actions and their consequences. In the current context, the aforementioned principle holds that the duty of decarbonization should impose obligations to oil companies in terms of emissions abatements that, while adequately stringent, do not penalize the entire society. Therefore, such obligations should be more rigorous in the socioeconomic systems that are least dependent on fossil fuels and have the greatest resources to address their wider societal implications as well as those with a greater ability to politically and technologically manage the low-carbon transition.

Sufficiency

The principle of sufficiency, by and large, demands that all agents should have enough to subsist above a certain threshold, below which it is impossible to have reasonable opportunities in life, that is, to have access to the basic environmental, social, and economic conditions required for a dignified life. Given the emphasized importance for social cohesion and protection against job losses associated with the duty of decarbonization, the achievement of such a principle should ensure secure livelihoods and stability for workers and communities within the fossil fuel industry's value chain.

* * *

Based on these two principles and taking into account the specifications about the operationalization and implementation of the duty of decarbonization for the top twenty oil and gas companies, it can be surmised that a morally sound process turning Big Oil into Big Green should be enacted and governed along the lines described below in order to be more feasible and to lessen possible negative socioeconomic implications.

First, the process should aim at preventing socioeconomic systems from being irremediably unsettled. Basically, any related actions, initiatives, and projects must cause the least possible damage to the fewest socioeconomic systems. Consistent with these considerations, one approach for incorporating and systematizing the clusters of practices and values that should guide Big Oil's decarbonization is fundamental: precaution. Given the extreme uncertainty that characterizes processes of decarbonization, precaution should, however, allow for contextual and emerging circumstances and elements.

Second, a major stumbling block to Big Oil's decarbonization processes and governance is that the companies lack the coordination qualities needed to achieve abatement requirements. To obviate or at least lessen this risk, decarbonization processes and governance should achieve the coordination required in a way that can defensibly be trusted in the long term (Buchanan and Keohane 2006). In short, these processes and governance must acquire and maintain legitimacy. Legitimacy is understood in this context as a normative property that favors the convergence of opinions on the need to endorse actions required by the duty of decarbonization.

Third, given the contentiousness and criticality of decarbonization, its processes and governance risk being appropriated, as chapter 7 outlines, by the oil complex and by elites, techno-scientific managers, bureaucrats, and profit-seeking investors, as the political economy of climate change and current climate politics evidence suggest. These composite challenges may encourage processes and forms of governance forged around the will of Big Oil or the other powerful groups, largely based on instrumental rationalities. To contrast this hazard, decarbonization arrangements should guarantee independence, as this normative property can actually lessen the possibility of vested interests coming into play and can also magnify the ability of processes and governance working in the public interest, even in the event of possible interference.

Fourth and finally, given the fact that the decarbonization of oil and gas companies is a costly matter and that oil-exporting countries would see a significant source of their revenues sacrificed, the related processes need to be financially supported. In this regard, it is worth recalling that chapter 9 suggests that to contrast the overall burden of decarbonization, the Fund for Oil Rectification should have two subsidiary units: the transition unit, to contribute to fund actions, initiatives, and projects to favor the low-carbon transition, and the workers and communities unit, which should support displaced workers and frontline communities. It also seems prudent to consider wealthier countries' support, channeled for instance through development aid, for the low-carbon transition in less capable/more vulnerable countries, such as certain petrostates (Armstrong 2020).

CONCLUSION

Despite its crucial role in promoting wealth and comfortable lifestyles worldwide and its capacity to shield its contribution to climate change, in the past few years the oil industry has quite abruptly found itself under a harsh spotlight, with accusations of responsibility for the climate crisis raining down on it. Major international oil companies (IOCs) have hurriedly responded to the scrutiny with a series of pledges, plans, and press releases aimed at clarifying their commitments, with varying degrees of ambition, to reduce carbon emissions.

BP's CEO Bernard Looney even admitted to understanding the mistrust the public may harbor with regard to the company's 2050 net-zero plan: "I get the sort of suspicion. But we are serious about this. This is in the interests of our company. It's not like we're trying to protect our existing business and get by. We are pivoting BP from being an international oil company that we've been for 111 years to becoming an integrated energy company" (Harder 2020). In fact, besides the alleged use of net-zero to muddy the waters over responsibility for climate change (ActionAid et al. 2020), a report published in September 2020 (OCI 2020) carried out a reality check against such claims and found that the actions and plans of oil giants—BP, Chevron, ENI, Equinor, ExxonMobil, Repsol, Shell, and TotalEnergies—are far from being aligned with the 1.5°C goal set by the Paris Agreement. More broadly, no IOCs decarbonized their business or cut down their involvement in fossil fuels in the period 2004–2019 (Green et al. 2020), and national oil companies (NOCs), besides some declarations of goodwill, are similarly getting on with business as usual.

In a nutshell, while the oil industry is expected to seriously engage with the climate crisis and claims to be doing so, the facts often contradict the fiction it peddles. This issue, of fast-growing concern in the academic and nonacademic debate on climate change, was the core object of the ten chapters of this book: the analysis of the role, responsibility, and duties of the oil and gas industry and of the consequent implications for the governance of the climate crisis.

IOCs and NOCs, by providing a deluge of fossil fuels to the global economy, are the driving force behind our current carbon-centric socioeconomic systems. Yet, they still remain somewhat overlooked in the climate discourse, an *elephant in the room* of the global climate debate and negotiations. It would appear that some run scared from daring to condemn the industry for its role in the climate crisis: for instance, after twenty-five years of United Nations (UN) negotiations, it was only during the 2019 COP 25 in Madrid that an official document dared to include the intractably hot *F-words, fossil fuels* (Abreu and Henn 2019). Similarly, the IPCC—which in its last report (IPCC 2021a) indeed sounded a "code red for humanity" and made evident that fossil fuel combustion contributed to 64 percent of the increase in human-caused carbon dioxide emissions since 1750, and to 86 percent of emission growth over the last 10 years— did not include neither in the forty-one pages of the report's *Summary for Policymakers* (IPCC 2021b) nor in the press release of its presentation the words *fossil fuels*. Rather, these documents timidly argued that "human activities" and "influence" are causing the current climate crisis, without specifying why this is happening and who should be held to account. The American sociologist Robert Brulle, who has long studied climate denial, said that UN climate science reports neglecting to mention the obstructive role of the fossil fuel lobby in the climate change narrative is "like trying to tell the story of Star Wars, but omitting Darth Vader" (Lo 2021).

Through their informed and self-advantageous choice to continue the exploration, production, refining, and distribution of fossil fuels after the associated risks became known to them, oil and gas companies have essentially imposed on the global socioeconomic system a carbon-intensive model of development. Rather than engaging in a large-scale search for alternatives and phasing out fossil fuels as warranted by the urgency of the climate crisis, something that their vast technical expertise and wealth would permit, these companies continued with their fossil fuel–dependent business models and behavior.

The main contribution of the book to the current scientific and policy debate on climate change is strengthening the acknowledgment that oil and gas companies are new central agents of climate ethics and policy. The book tries to draw attention to the fact that an extremely important group of agents—oil and gas companies—can be repositioned from the global villain to agents of change in the narrative of the climate crisis, as the title states, from Big Oil to Big Green. Their role in global climate governance should be consistent with the one they played in climate change along with states, individuals, and other agents. Broadening the perspective from states to oil and gas companies opens up new possibilities for them to become part of the solution rather than passive and profitable bystanders to continuous climate disruption. On the other hand, the social condemnation of fossil fuels and the prospect of escaping the current carbon lock-ins are greatly favored if oil and gas companies are recognized as crucial agents in climate change with specific duties.

This book builds its arguments starting from a moral issue: Big Oil violated the no-harm principle and thus bears responsibility and has duties of reparation and decarbonization to limit consequent harm. Moral approaches to the future and the natural world seem to have become more relevant because of the COVID-19 pandemic that struck the world in early 2020. Humanity worriedly ponders how it can curb the burden it places on future generations in an endangered planet. Therefore, the broad moral perspective taken by the book is now even more effective and is easier to digest by a general public and perhaps to follow by those who have issues at stake in its argument.

THE PANDEMIC, OIL, AND ENERGY

During the writing of this book, something exceptional and unforeseen—yet predictable and indeed predicted—abruptly and profoundly shook the world: the global health crisis caused by COVID-19. As the pandemic highlights the injustices and inequalities of our socioeconomic systems, at the same time it has severely affected the world economy and hit the oil industry.

A concluding chapter is not usually the place to throw in new arguments; however, *extraordinarily* disrupting circumstances such as those caused by the pandemic to the oil industry allow—probably require—an *extraordinary* conclusion. So in this spirit, the objective of this concluding chapter will be not so much to rake over the ground covered in the preceding chapters but rather to use them as a reference to better comprehend what happened to the oil industry within the world of energy in these unprecedented times and to briefly examine its potential role in a postpandemic world. A further and more general—albeit implicit—objective is to test, so to speak, the validity of the arguments of the book and of the implications it draws in the face of the overwhelming exogenous shock that the oil industry has undergone.

While concluding the revision of the book in September 2021, uncertainty still reigned: neither the origins and implications of the SARS-CoV-2 virus from which the pandemic originated nor how to properly address it, let alone the postpandemic recovery trajectory and pace, were clear. In fact, the recovery is uncertain. For instance, the International Energy Agency (IEA) *World Energy Outlook 2020* includes two energy demand scenarios dependent on the evolution of the pandemic: the *stated policies scenario,* which shows what would happen if governments continue with their current policies and the global economy recovers from the COVID-19 recession by 2021, and the *delayed recovery scenario,* which considers a recovery only from 2023 (IEA 2020e).

The three largest producers of greenhouse gases—the European Union, the United States, and China—envisage different trajectories that would push the planet in very different directions. Europe, which has set an emissions reduction goal for 2030 to at least 55 percent compared to 1990 and pledged to reach net-zero carbon emissions by 2050, outlines a green future, with a €672.5 ($770.6) billion recovery package aimed at transitioning its members away from fossil fuels by escalating wind and solar power, retrofitting old buildings, and investing in cleaner fuels such as (green) hydrogen. For supporting these commitments the European Commission has released in July 2021 a legislation package named *Fit for 55*. In 2020, renewables generated 38 percent of Europe's electricity (compared to 34.6 percent in 2019), for the first time overtaking fossil-fired generation, which went down to 37 percent (Agora Energiewende and Ember 2021).

In the United States, the Donald Trump administration seemed very attentive to the needs of the oil world: to sustain the economy through the pandemic, the US government gave the fossil fuel industry between $10.4 billion and $15.2 billion in federal direct economic relief, while the Federal Reserve fueled a lending boom of more than $93 billion in new bond issuances by oil and gas companies and purchased $432 million in oil and gas bonds from private investors (Wagner et al. 2020). However, in early 2021 the newly elected president, Joe Biden, signed a series of executive orders to address climate change, including a stop to new leasing of oil and gas developments on federal lands, and on March 31, 2021, proposed a $2.25 trillion infrastructure package (the eight-year *American Jobs Plan*) in which clean energy and climate action have a central role (Kolbert 2021). On Earth Day 2021 the Biden administration announced its commitment to abate carbon emissions by 50–52 percent below 2005 levels by 2030.

China plans to build new coal plants but has declined to set specific economic growth targets, a decision that can reduce the pressure on the country's industrial machine; indeed, surprisingly, Chinese premier X Jinping proclaimed at the 2020 UN General Assembly, in what is considered the most important announcement on addressing climate change in years, that his country would cut emissions to net zero by 2060. In December 2020 he pledged to abate carbon emissions per unit of economic output by over 65 percent by 2030 and boost the share of nonfossil fuels in energy consumption to roughly 25 percent by then, while in September 2021 announced the ending of Chinese involvement in the construction of coal-fired power plants overseas. All along, China's March 2021 Five-Year Plan championing the continued expansion of *clean coal* could lead to a substantial rise in greenhouse gas emissions and even hamper future climate targets (Normile 2021).

And yet, mysteriously, a report from the University of Oxford and the UN Environment Programme found that out of the $14.6 trillion pandemic-related fiscal rescue and recovery effort announced by the fifty largest economies in 2020, only 2.5 percent ($368 billion) was for green activities (Callaghan and Murdock 2021). At the same time, the 2020 *Production Gap Report* shows that to have a reasonable chance of avoiding 1.5°C or more of global heating, humanity needs to cut fossil fuel production by 6 percent per year until 2030 (SEI et al. 2020). Countries are instead planning and projecting an average annual increase of 2 percent, which in 2030 would more than double the production consistent with the 1.5°C limit.

According to the IEA (2020a), the pandemic is the biggest shock to the global energy system since World War II, six times more than the decline that followed in the wake of the 2008 financial crisis, and twice as much as the combined total of all previous reductions since the end of World War II. The world consumed 8.6 percent less oil in 2020 then the year before, but its demand could surpass prepandemic levels within a few years (IEA 2021d), and the demand for energy declined by 6 percent. This slump would mean a 5.8 percent reduction in global energy-related CO_2 emissions (IEA 2021a).

The COVID-19 pandemic uncompromisingly shows the enormous task our socioeconomic systems face in order to meaningfully address the climate crisis. Despite months-long lockdowns involving one-third of the world population, global carbon emissions decreased by only 7 percent in 2020 (Le Quéré et al. 2021), while daily global CO_2 emissions decreased by 17 percent by early April 2020 compared with the mean 2019 levels, almost half due to fewer car journeys (Le Quéré, Jackson, and Jones 2020a).[1] This was, however, only a temporary drop that did not reflect structural changes in the economic, transport, or energy systems. In fact, global CO_2 emissions had returned to prepandemic levels by the end of 2020 and even surpassed them in some major economies such as Brazil, China, and India, (IEA 2021a). In 2021 they are forecast to grow by the second-biggest annual rise in history (IEA 2021b).

At any rate, emissions should drop by the equivalent of a global lockdown roughly every two years for the next decade, hopefully in completely different ways, for the world to safely limit global heating (Le Quéré et al. 2021). Past postcrisis recoveries show that a low fossil fuel reboot makes more of an impact on climate than a profound yet brief emissions fall could. A serious engagement in a global low-carbon recovery could reduce carbon dioxide levels in the atmosphere by up to 19 parts per million by midcentury compared with a recovery that emphasizes fossil fuels (Hanna, Xu, and Victor 2020).

A glimmer of hope comes from renewable energy, which is rapidly deployed at scale (IEA 2021c). This source is expected to grow by nearly 4 percent globally, reaching almost 200 gigawatts in 2020, roughly 90 percent of all new generating-capacity additions (IEA 2020c). According to the IEA *World Energy Outlook 2020*, solar, the new *king of energy*, is already the "cheapest electricity in human history" and is going to set new records for deployment every year after 2022 (IEA 2020e). At the same time wind and solar capacity will exceed natural gas in 2023 and coal in 2024 (IEA 2020c). Astonishingly, humanity can already capture more than one hundred times the present global energy demand from solar and wind; the land required for solar panels alone to meet the global energy demand is only 0.3 percent of the global land area and *less* than the land currently used by fossil fuels (CTI 2021b).

In sum, amid the great uncertainty currently dominating our epoch, two contrasting recovery trajectories could emerge. On the one hand, the growth in renewables and greener energy seems unstoppable, yet cut-price oil beckons temptingly, with the possible consequence being a surge in the use of fossil fuels. Faith Birol, executive director of the IEA, presenting the *World Energy Outlook 2020*, argued that while the COVID-19 pandemic can reshape the future of energy, "the era of global oil demand growth will come to an end in the next decade, but without a large shift in government policies, there is no sign of a rapid decline. Based on today's policy settings, a global economic rebound would soon push oil demand back to pre-crisis levels" (IEA 2020f). In fact, the IEA's April 2021 estimates reveal that global oil demand in 2021 will be 5.7 million barrels per day above 2020. In April 2021 the US shale industry rose like a phoenix from the flames, and crude prices returned to the prepandemic level, making the business profitable again.

BIG OIL AND THE PANDEMIC

Within the broad picture illustrated above, the COVID-19 pandemic has wreaked havoc on the oil industry. What is certain is that the oil industry has survived many periods of hardship and in all likelihood will survive this one too, perhaps the harshest it has ever witnessed. Big Oil is nothing if not pragmatic and is therefore loath to *waste* a good crisis: it might, for instance, use the pandemic to push for legislation to criminalize protests against oil infrastructures or to increase its subsidization. Many observers are confident in the industry's ability to endure market volatility as oil majors had done in previous crises. Once the outbreak has been reined in, the global economy is expected to rebound and will likely be aching to quench its thirst for energy. It is still too early to clearly pinpoint who is going to quench this thirst and how. One crucial thing stands out starkly though: if low-carbon development strategies and policies are not rolled out in postpandemic recovery plans, the COVID-19 crisis will exacerbate climate change impacts if governments divert some of the resources for climate change to address the pandemic (Climate Action Tracker 2020).

A likely scenario in the postpandemic oil world is that the expected wave of bankruptcies among smaller fossil fuel companies will accelerate concentration of the industry in favor of the oil majors (Brower, Jacob, and Raval 2021).[2] According to Goldman Sachs, "Big Oil will consolidate the best assets in the industry and will shed the worst . . . when the industry emerges from this downturn, there will be fewer companies of higher asset quality" (Reuters 2020). Even if oil majors have declared multimillion- to multibillion-dollar losses in their earnings reports since the start of the pandemic, the *good prices* that the market expresses make it possible for them to buy more wells on the cheap and to amass more reserves.

On the other hand, it seems that oil majors' worst fears could be coming to pass: BP announced that it will be laying off ten thousand workers, wrote down the value of its oil and gas assets by up to $17.5 billion, lost $5.7 billion in 2020, and indeed may be forced to leave new fossil reserves in the ground.[3] Shell too has revealed that it is going to write down the value of its assets by more than $22 billion as the pandemic lowers demand for its oil and natural gas and its price forecasts; the Anglo-Dutch giant, which reported a $19.9 billion loss in 2020 while raising its dividend by almost 40 percent and launching $2 billion of share buybacks on July 2021 and announcing in September 2021 that it will distribute $7 billion of the proceeds from the Permian deal back to shareholders, will cut up to nine thousand jobs in an attempt to reposition itself in the energy transition. Both companies have stated that the accounting maneuvers were a response to not only the COVID-19 recession but also global efforts to tackle climate change. Italy's ENI has also declared the write-down of deferred tax assets of €3.5 ($4.0) billion, while ExxonMobil reported that it lost $22.4 billion in 2020, compared to a profit of $14.3 billion in 2019; $19.3 billion of the loss came from the write-down of assets. Altogether, oil majors have wrote off over $105 billion in their oil and gas assets in 2020 (IEA 2021d), whereas in the same year US companies got a $8.2 billion tax bailout and fired almost sixty thousand workers (Bailout Watch 2021).

Since oil majors are vertically integrated, they can offset their losses in upstream phases (e.g., production) with the lower cost of fuel inputs for their downstream operations. Shell, for instance, doubled its crude and refined products trading profits in 2020 compared to 2019 (Shell 2021). Larger oil and gas companies have reserves and assets distributed in all four corners of the globe, including the areas where oil and gas production is the cheapest. By doing this, Big Oil would switch to top-quality assets and become stronger and more powerful and its fossil fuel–related activities even more dangerous for the planet if humanity goes back to the prepandemic status quo centered on highly carbon-intensive socioeconomic systems.[4]

Indeed, it does appear that the industry is not planning to change its behavior or, as the prominent climate change activist and journalist Bill McKibben, in a clear reference to a certain US behemoth's famous advertising campaign,[5] provocatively affirmed in an editorial published in May 2016 in *The Guardian* that the oil industry "is not going to change its stripes" (McKibben 2016). Quite the contrary. Big Oil has already surreptitiously exploited the crisis by aggressively lobbying for massive bailouts and special privileges, as the $15.2 billion in direct economic relief from federal efforts under former president Trump and the $86 billion of offshore tax loopholes since 2017 (mentioned in chapter 6) as well as the substantial bond purchase mentioned above testify.

So, it would seem that most of the oil and gas companies' pledges about their virtuous low/zero-carbon future are misleading and misrepresent their willingness to change. None of the companies has committed to cut its oil and gas output over the next ten years, the simplest and most reliable indicator of actual change (CTI 2020a). Their stated *net-zero ambitions* are based on either capturing or offsetting these emissions with unproven technologies and reforestation at a questionable scale (Kusnetz 2020).

NOTHING HAS ACTUALLY CHANGED

After this dizzying journey into the mysteries of an invisible enemy and the reactions of a powerful yet unpredictable industrial behemoth, it is necessary to take stock. A few hard-nosed facts may help to see where the oil industry is and intends to be in this brave new world. These emblematic facts—all subsequent to the start of the COVID-19 outbreak in early 2020—offer a bird's-eye view as to whether or not Big Oil is indeed striving to shed its skin and become Big Green Energy.

In the September 17, 2020, exploration plans Shell filed with the State of Alaska, the company unveiled its intention to resume oil and gas exploration in Alaskan Arctic offshore waters for the first time since 2015 (Mower and Bradner 2020); the company also funded lobbying to push for a rule to block banks' policies against lending for Arctic drilling and coal mining (Barratt and Kaufman 2021). The same goes for the French IOC TotalEnergies, part of the group of the net-zero 2050 pledge in the virtuous Europe of the Green Deal, that is projected to increase its fossil fuel production by 12 percent in 2030 (McKibben 2020b).

In October 2020 Bloomberg, after careful analysis of ExxonMobil's internal documents, announced that the Texan giant has been planning to increase annual carbon dioxide emissions—failing to disclose these estimates to investors—by as much as the output of the entire nation of Greece. Amazingly, this figure refers only to carbon emissions from direct operations (i.e., scope 1 emissions) caused by the seven-year investment plan adopted in 2018 by CEO Darren Woods (Crowley and Rathi 2020), whereas, as repeatedly pointed out, the bulk of emissions comes from downstream combustion of the fossil fuels sold to the global economy (i.e., scope 3 emissions). In the meantime the company is spending millions on social media advertising to rally support to fight for oil and gas interests at every level of government (MacDonald 2020).

In 2020, the fossil fuel industry heavily lobbied key members of the European Commission in charge of the European Green Deal to push for the inclusion of solutions that allow the industry to maintain its business model, based on the extraction and production of fossil fuels, and to control the energy transition so as not to turn off the golden tap (Corporate Europe Observatory, Food and Water Action Europe, and Re:Common 2020).

Despite the *"build back greener"* plan launched by the British government during the COVID-19 pandemic, UK ministers met representatives from oil and gas companies including BP, ExxonMobil, and Shell 149 times between April and June 2020, while they met renewable energy producers just 17 times over the same period (Cooke 2020).

NOCs, which currently produce about two-thirds of the world's oil and gas and own about 90 percent of reserves, plan to invest about $400 billion in the next decade in oil and gas projects that—tellingly—can only break even if humanity exceeds the global carbon budget and allows the global temperature to rise more than 2°C (Manley and Heller 2021).

The list goes on. So no, it does not seem that Big Oil is turning over a new leaf. Therefore, to dutifully answer the question pondered at the beginning of this concluding chapter, yes, the normative-theoretical and positive-empirical arguments developed in the book are still valid. Despite—or even more so because of—the destruction of livelihoods, the loss of life at a devastating scale, and the disruptions to manufacturing and supply chains that the pandemic has caused, Big Oil should still meet its duties of reparation and decarbonization, and agents of destabilization should be even more proactive in favoring their achievement.

In sum, although the disruptive potential on the industry of the COVID-19 crisis still remains to be seen, a number of signals testify that the old order will fight back, as has so often been the case in history. A chorus of reactionary voices is already trying to push back onto center stage the *old normal*, a way of life that gravitated around the oil industry. The optimistic view that COVID-19 would heighten awareness of the other risks humanity faces and of the value of precautionary action is being edged out of the picture. The threat of climate change simply does not seem immediate or palpable enough to change the dismissive attitude shown so far.

THE POTENTIAL OF BIG OIL IN THE POSTPANDEMIC RECOVERY

The climate and COVID-19 crises bear fundamental similarities: both are mass global threats, there is clear science to address them, and both clarify how standard *old normal* behavior can result in catastrophic outcomes. Additionally, awareness of the strong correlation between health and climate issues (Watts et al. 2021) as well as between COVID-19 and exposure to hazardous air pollutants, largely derived from the combustion of fossil fuels (Petroni et al. 2020), is fast growing. For instance, in the first half of the 2020 phase of the pandemic, an incredible 78 percent of pandemic-related deaths across sixty-six administrative regions in France, Germany, Italy, and Spain occurred in the five most polluted ones (Ogen 2020). By the same token, increased exposure to air pollution contributes to the disproportionate impact that COVID-19 has on Black communities and more generally on racial minorities in the United States.

The ongoing COVID-19 crisis is merely a dress rehearsal for possible climate catastrophe. Unfortunately, there is no equivalent to Big Pharma scrambling around desperately to find a vaccine for climate change, and the current social and economic disruptions will pale in the face of a global climate crisis in full bloom. Many of the more than one hundred disasters that impacted over fifty million people in 2020 alone are related to climate change (IFRC 2020).

The pandemic has engendered unprecedented drastic and costly measures to obviate the threat. Major financial and policy decisions made now will shape the global economy over the next decade, precisely the period in which humanity must halve emissions; a swift and coherent shift to a low-carbon future that prioritizes human health over profit will simultaneously flatten both emissions and virus curves. Once urgent health and social protection measures have been deployed, inclusive recovery programs should prioritize low-carbon socioeconomic systems, thereby reducing the impact that current fossil fuel–based models have on health care systems due to illnesses aggravated by poor air quality.

Climate and health should not be playing off each other to attract resources; tackling the climate crisis and reducing the risk of pandemic-causing pathologies are parallel long-term challenges that require unshackling our socioeconomic structures from fossil fuels by involving the oil industry in a sustainable and systemic postpandemic recovery. Humanity must aim to build socioeconomic systems predominantly fueled by renewable energy, and the oil and gas industry—the lifeblood of the current carbon intensive world and the agent with the greatest responsibility for the climate crisis and therefore an obligation to make amends—provides an entry point to drastically remodel these systems.

Let us use this closing argument to return to the expression *Big Oil*: the term is usually employed in a disparaging way by detractors, underlining the massive economic and political clout these companies wield, not least due to their lobbying influence, and the ironclad grip their products hold on industrial society. By baking reliance on their products into the recipe and practicing the idolatry of growth, Big Oil has created a Ponzi scheme that has been running for the past 150 years whereby future economic and environmental stability is sacrificed for the immediate riches of the few for a couple more generations. That placement of the word *Big* before any number of industries (Big Meat, Big Pharma, Big Tech, Big Tobacco, take your pick) makes the business seem ominous, impersonal, a faceless corporate entity that has sold the soul of innovative industriousness to an economic and political system run on the concept of pure growth. But there's the rub: has Big Oil become so big that it can dictate its terms thanks to a system that sustains it, or is it sustaining the system with its political financing and lobbying influence? It is a chicken-and-egg scenario, one that must be tackled on both fronts, as this book argues.

And so on to the hoped-for transition of the industry, in a kind of reimaged Midas touch, whereby if Big Oil were to embrace the book's recommendations to decarbonize, everything it touches would turn to green. Of course, Big Green—considering the aforementioned interpretation of the adjective *Big* in industry—seems a juxtaposition of contradictory terms—*big* is bad, *green* is good—an oxymoron that could prove to epitomize the moral dilemmas with regard to industry in our times in so many ways. And yet the entrepreneur, as noted by Joseph Schumpeter in his analysis of the business cycle, is the driver of technological innovation to benefit society in the "process of industrial mutation that continuously revolutionizes the economic structure from within, incessantly destroying the old one, incessantly creating a new one" (Schumpeter 1942, 83).

"They have to avoid fighting gravity," one speaker told delegates at perhaps the most influential annual energy conference, CERAWeek, held remotely because of the pandemic in March 2021, in reference to how the oil industry must stop fighting the inevitable and transition to green energy or risk not being at the table anymore but being on the menu, so to speak. Another speaker was John Kerry, President Biden's special envoy on climate change: "And I think the fossil fuel industry could clearly do a lot more to transition into being a full-fledged energy company that is embracing some of these new technologies," Kerry said. "There has been enormous growth in investment, in longer-term speculation investment and I think it's a clear 'why.' Predictions are that by 2050 you're going to have about 6 trillion dollars a year of economic transfer taking place in the clean energy technology sector. It's the market of the future" (US Department of State 2021). With figures like that being bandied about, it could encourage Big Oil from taking baby steps to taking leaps and bounds to becoming Big Green.

☆　☆　☆

"Can . . . the leopard change his spots? Then may ye also do good, that are accustomed to do evil."

This passage from the King James Bible contradicts the idiom that has fallen into popular usage, that an *evil* entrenched nature cannot be modified to *do good*. Indeed, the extended concept exemplifies the ultimate thesis of this book: that moral wrongdoings can, to some extent, be righted.

From Big Oil to Big Green does not adopt a cry to arms that the oil industry be done away with altogether, a crude (no pun intended) and improbable outcome. Instead, this book promotes the idea that a resolvable balance needs to be struck between the economic rights in the capitalist society we—like it or not—live in and other more universal rights, prioritizing without taking a sword and executing an industry that is one of the bedrocks for our global system.

NOTES

Introduction

1. Throughout the book, *dollar* refers to US currency.

Chapter 1

1. The permit for the Keystone XL pipeline was revoked by US president Joe Biden's executive order of January 20, 2021. On June 9, 2021, the Canadian TC Energy Corporation, owner of Keystone XL pipeline, terminated the project.

2. For instance, the Climate Investigation Center, DeSmog, and the Pulitzer prize–winning Inside Climate News.

Chapter 2

1. Data taken from the companies' *Annual Reports* and *Forms 20-F* to the US Securities and Exchange Commission 2015, 2016, 2017, 2018, and 2019.

Chapter 3

1. See, for instance, *The People's Demands for Climate Justice*: https://www.peoplesdemands.org/.

2. Chapter 5 provides a thorough rebuttal of Big Oil's *"blame the consumer"* strategy.

Chapter 5

1. For the Repsol announcement, see Repsol (2019).

2. For the Shell announcement, see Shell (2020b).

Chapter 6

1. For details, see Sabin Center for Climate Change Law (2021).

2. It should be noted, however, that this enormous figure is much broader than actual cash transfers from governments to the fossil fuel industry. The latter coincide with a narrower understanding of subsidies—closer, in fact, to the commonsense definition—according to which they amount to $296 billion for 2017 (Coady et al. 2019); including consumer-based fossil fuel subsidies, the amount would be $372 billion for 2018 (Bridle et al. 2019). In any case, the huge discrepancy between the narrow and broad understandings of fossil fuel subsidies depends on the fact that the first meaning—defined in the IMF report (Coady et al. 2019) as *posttax subsidy*, as opposed to the second, known as *pretax subsidy*—reflects the difference between *actual consumer fuel prices* and the full societal and environmental costs of a fuel, that is, the externalities associated with fossil fuel combustion. These externalities—in general, environmental pollution and its social repercussions—are obviously very large: posttax subsidies in the report amount, in fact, to roughly $ 4.9 trillion, or 94 percent of the total of $ 5.2 trillion. At any rate a handful of major IOCs—including Chevron, ConocoPhillips, and ExxonMobil—have benefited $86 billion's worth of subsidies in the form of offshore tax loopholes under the Trump Administration's 2017 *Tax Cuts and Jobs Act* (Friends of the Earth, Bailout Watch, and Oxfam 2021).

Chapter 8

1. For continually updated databases on national-level climate change laws, policies, and climate litigation cases globally, see Grantham Research Institute (n.d.) and Columbia University (n.d.).

Chapter 9

1. This mechanism, though, will be financed over the 2021–2027 period through the Just Transition Fund (initially equipped with €7.5 [$8.7] billion and then increased to €17.5 [$20.0] in July 2020), €45 ($51.6) billion of mobilized investments, and a public-sector loan facility backed by the European Investment Bank amounting to €25–30 ($28.7–34.4) billion (European Commission 2020).

Chapter 10

1. Measured through the 2020 GDP per capita calculated by the International Monetary Fund (IMF 2021).

Conclusion

1. According to a June 2020 update of this study (Le Quéré et al. 2020b), the 2020 decrease in global emissions compared to 2019 is only 5 percent. The authors claimed that although they expected a rebound (mostly coming from the transport sector), its rapidity was a big surprise.

2. According to Reuters, the top executives of ExxonMobil and Chevron held preliminary talks in early 2020 to explore the possibility of merging the two companies to better resist the challenges of the pandemic (Spector 2021).

3. However, in the first and second quarter of 2021 BP recorded, respectively, $2.6 and $2.8 billion profit while in the latter quarter carried out a $500 million share buyback.

4. The environmentalist and campaigner Bill McKibben argues instead that the COVID-19 pandemic has reduced the industry's power. "The coronavirus crisis has both obscured and illuminated one of the most seismic developments on our planet in many decades: I think it's now clear that the power of the fossil-fuel industry has decisively passed its zenith" (McKibben 2020a).

5. The reference is to the Esso gasoline "Put a Tiger in Your Tank" ads from the 1960s.

REFERENCES

Abatzoglou, J. T., and A. P. Williams. 2016. "Impact of Anthropogenic Climate Change on Wildfire across Western US Forests." *Proceedings of the National Academy of Sciences* 113 (42): 11770–11775.

Abbasi, K., and F. Goodlee. 2020. "Investing in Humanity: The BMJ's Divestment Campaign." *BMJ* 368. https://www.bmj.com/content/368/bmj.m167.

Abbott, K. W. 2012. "The Transnational Regime Complex for Climate Change." *Environment and Planning C: Government and Policy* 30 (4): 571–590.

Abreu, C., and J. Henn. 2019. "Finally Saying the F-Words at UN Climate Talks." *ClimateHome News*, December 16, https://www.climatechangenews.com/2019/12/16/finally-saying-f-words-un-climate-talks/.

Achakulwisut, P., P. Erickson, and D. Koplow. 2021. "Effect of Subsidies and Regulatory Exemptions on 2020–2030 Oil and Gas Production and Profits in the United States." *Environmental Research Letters* 16 (8): 084023.

ActionAid and Climate Justice Organisations. 2020. *NOT ZERO: How 'Net Zero' Targets Disguise Climate Inaction*, October, https://actionaid.org/sites/default/files/publications/NOT%20ZERO_Joint%20Technical%20Briefing.pdf.

Agora Energiewende and Ember. 2021. *The European Power Sector in 2020: Up-to-Date Analysis on the Electricity Transition*. https://ember-climate.org/wp-content/uploads/2021/01/Report-European-Power-Sector-in-2020.pdf.

Aklin, M., and M. Mildenberger. 2020. "Prisoners of the Wrong Dilemma: Why Distributive Conflict, Not Collective Action, Characterizes the Politics of Climate Change." *Global Environmental Politics* 20 (4): 4–27.

Alderman, L. 2019. "Lagarde Vows to Put Climate Change on the E.C.B.'s Agenda." *New York Times*, September 4, https://www.nytimes.com/2019/09/04/business/climate-change-ecb-lagarde.html.

Almiron, N., M. Boykoff, M. Narberhaus, and F. Heras. 2020. "Dominant Counter-Frames in Influential Climate Contrarian European Think Tanks." *Climatic Change* 162: 2003–2020.

Amanze-Nwachuku, C. 2007. "Nigeria: NNPC's Oil and Gas Assets Base Hit $56bn." All Africa, November 5, https://allafrica.com/stories/200711050852.html.

Ambrose, J. 2019. "BlackRock Lost $90bn Investing in Fossil Fuel Companies, Report Finds." *The Guardian*, July 21, https://www.theguardian.com/environment/2019/jul/31/blackrock-lost-90bn-investing-in-fossil-fuel-companies-report-finds.

Ambrose, J. 2021. "Shell Calls on Investors to Vote for Its New Climate Strategy." *The Guardian*, April 15, https://www.theguardian.com/business/2021/apr/15/shell-calls-on-investors-to-vote-for-its-new-climate-strategy.

Anderson, K., and G. Peters. 2016. "The Trouble with Negative Emissions." *Science* 354 (6309): 182–183.

Ansar, A., B. L. Caldecott, and J. Tilbury. 2013. *Stranded Assets and the Fossil Fuel Divestment Campaign: What Does Divestment Mean for the Valuation of Fossil Fuel Assets?* Smith School of Enterprise and the Environment. Oxford: University of Oxford, Stranded Asset Programme.

Apolitical. n.d. "Climate 100: The World's Most Influential People in Climate Policy." https://apolitical.co/lists/most-influential-climate-100/. Accessed on September 25, 2021.

Armstrong, C. 2020. "Decarbonisation and World Poverty: A Just Transition for Fossil Fuel Exporting Countries?" *Political Studies* 68 (3): 671–688.

Arnold, D. G., and K. Bustos. 2005. "Business, Ethics, and Global Climate Change." *Business & Professional Ethics Journal* 24 (1–2): 103–130.

Asheim, G. B., T. Fæhn, K. Nyborg, M. Greaker, C. Hagem, B. Harstad, and K. E. Rosendahl. 2019. "The Case for a Supply-Side Climate Treaty." *Science* 365 (6451): 325–327.

Atkin, E. 2019. "Introducing: The Fossil Fuel Ad Anthology." *Heated*, December 13, https://heated.world/p/introducing-the-fossil-fuel-ad-anthology.

Attari, S. Z., D. H. Krantz, and E. U. Weber. 2019. "Climate Change Communicators' Carbon Footprints Affect Their Audience's Policy Support." *Climatic Change* 154 (3–4): 529–545.

Avelino, F., and J. Rotmans. 2009. "Power in Transition: An Interdisciplinary Framework to Study Power in Relation to Structural Change." *European Journal of Social Theory* 12 (4): 543–569.

Baatz, C. 2017. "Compensating Climate Change Victims in Developing Countries—Justification and Realization." PhD diss., University of Greifswald.

Baatz, C. 2018. "Climate Adaptation Finance and Justice: A Criteria-Based Assessment of Policy Instruments." *Analyse & Kritik* 40 (1): 73–106.

Baggini, J. 2018. "Reparations for Slavery Are Not about Punishing Children for Parents' Sins." *The Guardian*, November 30, https://www.theguardian.com /commentisfree/2018/nov/30/reparations-slavery-sins-parent-child-reparative justice-holocaust.

Bailout Watch. 2021. *Fossil Fuel Firms Slashed Nearly 60,000 Jobs in 2020 While Pocketing $8.2 Billion Tax Bailout.* April 2, https://bailoutwatch.org/analysis/fossil fuel-firmsslashed-nearly-60000-jobs-in-2020.

Baker, S. 2020. "Denmark's MP Pension Divests 24 Oil Company Holdings." *Pensions&Investment*, June 9, https://www.pionline.com/esg/denmarks-mp pension-divests-24-oil-company-holdings.

Banerjee, N. 2015. "Exxon's Oil Industry Peers Knew about Climate Dangers in the 1970s, Too." *Inside Climate News*, December 22, https://insideclimatenews org/news/22122015/exxon-mobil-oil-industry-peers-knew-about-climate-change dangers-1970s-american-petroleum-institute-api-shell-chevron-texaco.

Barratt, L., and A. C. Kaufman. 2021. "Despite Its Ledges, Shell Funded Anti-Climate Lobbying Last Year." Huffington Post, February 25, https://www.huffpost com/entry/shell-climate-lobbying-fossil-fuels_n_602d4530c5b66dfc101baac1.

Barry, C., and G. Øverland. 2016. *Responding to Global Poverty: Harm, Responsibility, and Agency.* Cambridge: Cambridge University Press.

Benoit, D. 2020. "JPMorgan Pledges to Push Clients to Align with Paris Climate Agreement." *Wall Street Journal*, October 6, https://www.wsj.com/articles/jpmorgan -pledges-to-push-clients-to-align-with-paris-climate-agreement-11602018245.

Bergquist, M., A. Nilsson, and W. Schultz. 2019. "Experiencing a Severe Weather Event Increases Concerns about Climate Change." *Frontiers in Psychology* 10: 220.

Bernstein, S., and M. Hoffmann. 2019. "Climate Politics, Metaphors and the Fractal Carbon Trap." *Nature Climate Change* 9: 919–925.

Binz, C., S. Harris-Lovett, M. Kiparsky, D. L. Sedlak, and B. Truffer. 2016. "The Thorny Road to Technology Legitimation—Institutional Work for Potable Water Reuse in California." *Technological Forecasting and Social Change* 103: 249–263.

Black-Kalinsky, C. 2016. "My Father Warned Exxon about Climate Change in the 1970s. They Didn't Listen." *The Guardian*, May 25, http://www.theguardian .com/commentisfree/2016/may/25/exxon-climate-change-greenhouse-gasses ?CMP=share_btn_link.

BlackRock Investment Institute. 2019. *Getting Physical. Scenario Analysis for Assessing Climate-Related Risks*. New York. https://www.blackrock.com/us/individual /literature/whitepaper/bii-physical-climate-risks-april-2019.pdf.

Blondeel, M. 2019. "Taking Away a 'Social Licence': Neo-Gramscian Perspectives on an International Fossil Fuel Divestment Norm." *Global Transitions* 1: 200–209.

BloombergNEF. 2021. "Integrated European Majors Lead on Preparedness for a Low-Carbon World among 39 Global O&G Companies." March 24, https://about .bnef.com/blog/integrated-european-majors-lead-on-preparedness-for-a-low -carbon-world-among-39-global-og-companies/.

Bonneuil, C., P. L. Choquet, and B. Franta. 2021. "Early Warnings and Emerging Accountability: Total's Responses to Global Warming, 1971–2021." *Global Environmental Change*, 102386.

Bordoff, J. 2020. "Everything You Think about the Geopolitics of Climate Change Is Wrong." *Foreign Policy*, October 5, https://foreignpolicy.com/2020/10/05/climate -geopolitics-petrostates-russia-china/.

Boudet, H., L. Giordono, C. Zanocco, H. Satein, and H. Whitley. 2020. "Event Attribution and Partisanship Shape Local Discussion of Climate Change after Extreme Weather." *Nature Climate Change* 10 (1): 69–76.

Bousso, R., and D. Zhdannikov. 2019. "Exclusive: No Choice but to Invest in Oil, Shell CEO Says." *Reuters*, October 24, https://www.reuters.com/article/us-shell-climate -exclusive/exclusive-no-choice-but-to-invest-in-oil-shell-ceo-says-idUSKBN1WT2JL.

Bowie, N. E. 2013. "Morality, Money, and Motor Cars Revisited." In *Business Ethics in the 21st Century*, 131–146. Dordrecht: Springer.

BP. 2020a. *BP Creative Workshop: Briefing Document*. January, https://assets .documentcloud.org/documents/20073850/bp-creative-workshop-v3-no-film.pdf.

BP. 2020b. *BP Energy Outlook*. https://www.bp.com/en/global/corporate/energy -economics/energy-outlook.html.

Bradshaw, C. J., P. R. Ehrlich, A. Beattie, G. Ceballos, E. Crist, J. Diamond, and D. T. Blumstein. 2021. "Underestimating the Challenges of Avoiding a Ghastly Future." *Frontiers in Conservation Science*, January 13, https:doi:10.3389/fcosc.2020.615419.

Brannon, H. R., A. C. Daughtry, D. Perry, W. W. Whitaker, and M. Williams. 1957. "Radiocarbon Evidence on the Dilution of Atmospheric and Oceanic Carbon by Carbon from Fossil Fuels." *Eos, Transactions American Geophysical Union* 38 (5): 643–650.

Braungardt, S., J. Bergh, and T. Dunlop. 2019. "Fossil Fuel Divestment and Climate Change: Reviewing Contested Arguments." *Energy Research & Social Science* 50: 191–200.

Bridge, G., and P. Le Billon. 2017. *Oil.* 2nd ed. Cambridge, UK: Polity.

Bridle, R., S. Sharma, M. Mostafa, and A. Geddesidle. 2019. *Fossil Fuel to Clean Energy Subsidy Swaps: How to Pay for an Energy Revolution.* Geneva: IISD. https://www.iisd.org/library/fossil-fuel-clean-energy-subsidy-swap.

Brooks, N., W. N. Adger, and P. M. Kelly. 2005. "The Determinants of Vulnerability and Adaptive Capacity at the National Level and the Implications for Adaptation." *Global Environmental Change* 15: 151–163.

Broome, J. 2019. "Against Denialism." *The Monist* 102 (1): 110–129.

Brower, D., J. Jacob, and A. Raval. 2021. "Big Oil's Huge Losses Raise Prospect of Mega Mergers." *Financial Times*, February 8, https://www.ft.com/content/79e64c9-a264-4a06-81f-b1d07d90fbdo.

Brown, G. 2019. "Goldman Sachs Leads among US Banks with Commitment Not to Fund Arctic Drilling." *Sierra Club*, December 15, https://www.sierraclub.org/press-releases/2019/12/goldman-sachs-leads-among-us-banks-commitment-not-fund-arctic-drilling.

Browne, J. 1997. *"Climate Change Speech by John Browne, Group Chief Executive, British Petroleum (BPAmerica),* Stanford University, 19 May 1997." DocumentCloud, http://www.documentcloud.org/documents/2623268-bp-john-browne-stanford-997-climate-change-speech.html.

Brulle, R. J. 2018. "The Climate Lobby: A Sectoral Analysis of Lobbying Spending on Climate Change in the USA." *Climatic Change* 149 (3–4): 289–303.

Buchanan, A., and R. O. Keohane. 2006. "The Legitimacy of Global Governance Institutions." *Ethics & International Affairs* 20 (4): 405–437.

Buchanan, A., and R. Powell. 2015. "The Limits of Evolutionary Explanations of Morality and Their Implications for Moral Progress." *Ethics* 126 (1): 37–67.

Buchner, B., A. Clark, A. Falconer, R. Macquarie, C. Meattle, R. Tolentini, and C. Wetherbee. 2019. *Global Landscape of Climate Finance 2019.* London: Climate Policy Initiative. https://climatepolicyinitiative.org/publication/global-landscape-of-climate-finance-2019/.

Buckley, T. 2019. "Over 100 Global Financial Institutions Are Exiting Coal, with More to Come." *Institute for Energy Economics and Financial Analysis,* February 27. http://ieefa.org/wp-content/uploads/2019/02/IEEFA-Report_100-and-counting_Coal-Exit_Feb-2019.pdf.

Burger, M., R. Horton, and J. Wentz. 2020. "The Law and Science of Climate Change Attribution." *Columbia Journal of Environmental Law* 45: 1–185.

Burger, M., and J. Wentz. 2018. "Holding Fossil Fuel Companies Accountable for Their Contribution to Climate Change: Where Does the Law Stand?" *Bulletin of the Atomic Scientists* 74 (6): 397–403.

Caesar, L., G. D. McCarthy, D. J. R. Thornalley, N. Cahill, and S. Rahmstorf. 2021. "Current Atlantic Meridional Overturning Circulation Weakest in Last Millennium." *Nature Geoscience* 14: 118–120.

Calcuttawala, Z. 2017. "The Secret Wealth of the World's Richest Oil Billionaires." Oilprice.com, March 1, https://oilprice.com/Energy/Energy-General/The-Secret-Wealth-Of-The-Worlds-Richest-Oil-Billionaires.html.

Callaghan, B. J., and E. Murdock. 2021. *Are We Building Back Better? Evidence from 2020 and Pathways for Inclusive Green Recovery Spending.* UNEP and University of Oxford. https://wedocs.unep.org/bitstream/handle/20.500.11822/35281/AWBBB.pdf.

Callendar, G. S. 1938. "The Artificial Production of Carbon Dioxide and Its Influence on Temperature." *Quarterly Journal of the Royal Meteorological Society* 64 (275): 223–240.

Caney, S. 2005. "Cosmopolitan Justice, Responsibility, and Global Climate Change." *Leiden Journal of International Law* 18: 747–775.

Caney, S. 2010. "Climate Change and the Duties of the Advantaged." *Critical Review of International Social and Political Philosophy* 13 (1): 203–228.

Caney, S. 2014. "Two Kinds of Climate Justice: Avoiding Harm and Sharing Burdens." *Journal of Political Philosophy* 22 (2): 125–149.

Caney, S. 2016a. "Climate Change and Non-Ideal Theory: Six Ways of Responding to Noncompliance." In *Climate Justice in a Non-Ideal World,* ed. C. Heyward and D. Roser, 21–42. Oxford: Oxford University Press.

Caney, S. 2016b. "The Struggle for Climate Justice in a Non-Ideal World." *Midwest Studies In Philosophy* 40 (1): 9–26.

Cann, H. W., and L. Raymond. 2018. "Does Climate Denialism Still Matter? The Prevalence of Alternative Frames in Opposition to Climate Policy." *Environmental Politics* 27 (3): 433–454.

Carney, M., F. Galhau, and F. Elderson. 2019. "The Financial Sector Must Be at the Heart of Tackling Climate Change." *The Guardian,* April 17, https://www.theguardian.com/commentisfree/2019/apr/17/the-financial-sector-must-be-at-the-heart-of-tackling-climate-change.

Cavendish, C. 2019. "Carbon Offset Gold Rush Is Distracting Us from Climate Change." *Financial Times,* https://www.ft.com/content/e2000050-0c7f-11ea-bb52-34c8d9dc6d84.

CDP. 2017. *The Carbon Majors Database—CDP Carbon Majors Report 2017.* https://www.cdp.net/en//articles/media/new-report-shows-just-100-companies-are-source-of-over-70-of-emissions.

CDP. 2018. *Beyond the Cycle.* https://www.cdp.net/en/reports/downloads/3858.

Chubb. 2019. *Chubb Limited Annual Report 2018.* http://s1.q4cdn.com/677769242/files/doc_financials/2019/AGM/Chubb-Limited-2018-Annual-Report.pdf.

CIEL (Center for International Environmental Law). 2017. *Smoke and Fumes: The Legal and Evidentiary Basis for Holding Big Oil Accountable for the Climate Crisis.* Washington, DC: CIEL.

Climate Action Tracker. 2020. "A Government Roadmap for Addressing the Climate and Post COVID-19 Economic Crises." *Climate Action Tracker Update, April,* https://climateactiontracker.org/documents/706/CAT_2020-04-27_Briefing_COVID19_Apr2020.pdf.

Coady, D., I. Parry, N.-P. Le, and B. Shang. 2019. *Global Fossil Fuel Subsidies Remain Large: An Update Based on Country-Level Estimates.* Washington, DC: IMF. https://www.imf.org/en/Publications/WP/Issues/2019/05/02/Global-Fossil-Fuel-Subsidies-Remain-Large-An-Update-Based-on-Country-Level-Estimates-46509.

Cohen, R. W. 1982. *Communication to A. M. Natkin, Office of Science and Technology, Exxon.* https://www.climatefiles.com/exxonmobil/1982-exxon-memo-summarizing-climate-modeling-and-co2-greenhouse-effect-research/.

Coleman, N. 2020. *"New Equinor CEO Vows Net Zero Emissions from Operations and Products by 2050."* S&P Global Platts, November 20, https://www.spglobal

.com/platts/en/market-insights/latest-news/coal/110220-new-equinor-ceo-vows
-net-zero-emissions-from-operations-and-products-by-2050.

Colgan, J., J. F. Green, and T. Hale. 2020. *Asset Revaluation and the Existential Politics of Climate Change.* Forthcoming: International Organization.

Columbia University. n.d. "Climate Change Litigation Databases." http://clima tecasechart.com/. Accessed on September 25, 2021.

Cook, J., D. Nuccitelli, S. A. Green, M. Richardson, B. Winkler, R. Painting, and A. Skuce. 2013. "Quantifying the Consensus on Anthropogenic Global Warming in the Scientific Literature." *Environmental Research Letters* 8 (2): 024024.

Cook, J., N. Oreskes, P. T. Doran, W. R. Anderegg, B. Verheggen, E. W. Maibach and D. Nuccitelli. 2016. "Consensus on Consensus: A Synthesis of Consensus Estimates on Human-Caused Global Warming." *Environmental Research Letters* 11 (4): 048002.

Cooke, P. 2020. "*Ministers Met with Fossil Fuel Producers 149 Times While Planning Green Recovery.*" DeSmog, November 10, https://www.desmog.co.uk/2020/11/10 /beis-government-ministers-fossil-fuels-meetings-covid-green-recovery.

Cooke, P. 2021. "Shell's 'Delusional' Net Zero Strategy Commits $8 Billion to Fossil Fuels." *DeSmog,* February 11, https://www.desmog.co.uk/2021/02/11/shell -net-zero-oil-fossil-fuels.

Cooke, P., R. Sherrington, and M. Hope. 2021. "*Revealed: The Climate-Conflicted Directors Leading the World's Top Banks.*" DeSmog, April 6, https://www.desmo gblog.com/2021/04/06/revealed-climate-conflicted-directors-leading-the-worlds -top-banks/.

Corkery, M. 2019. "Giant Factory Rises to Make a Product Filling Up the World: Plastic." *New York Times,* August 12, https://www.nytimes.com/2019/08/12/busi ness/energy-environment/plastics-shell-pennsylvania-plant.html.

Corporate Europe Observatory. 2020. "A Grey Deal? Fossil Fuel Fingerprints on the European Green Deal." *Corporate Europe Observatory,* July 7, https://corpora teeurope.org/en/a-grey-deal.

Corporate Europe Observatory, Food and Water Action Europe, and Re:Common. 2020. *The Hydrogen Hype: Gas Industry Fairy Tale or Climate Horror Story?* Brussels. https://corporateeurope.org/sites/default/files/2020-12/hydrogen-report-web-final_0 .pdf.

Crippa, M., G. Oreggioni, D. Guizzardi, M. Muntean, E. Schaaf, E. Lo Vullo, E. Solazzo, F. Monforti-Ferrario, J. G. J. Olivier, and E. Vignati. 2019. "Fossil CO2 and GHG Emissions of All World Countries—2019 Report." Luxembourg: Office of the European Union Publications. https://doi:10.2760/687800.

Crowley, K. 2019. "*Exxon Boosts Spending to $32 Billion, Raises 2025 Profit Target.*" Bloomberg Business, March 6, https://www.bloomberg.com/news/articles/2019-03-06/exxon-targets-32-billion-in-annual-spending-on-drilling-plants.

Crowley, K. 2021. "*Exxon to Hold Output at 20-Year Low to Address Debt, Emission.*" Bloomberg Business, March 3, https://www.bloomberg.com/news/articles/2021-03-03/exxon-outlines-plan-to-boost-returns-and-grow-its-dividend.

Crowley, K., and A. Rathi. 2020. "Exxon's Plan for Surging Carbon Emissions Revealed in Leaked Documents." *Bloomberg Energy & Science*, October 5, https://www.bloomberg.com/news/articles/2020-10-05/exxon-carbon-emissions-and-climate-leaked-plans-reveal-rising-co2-output.

CTI (Carbon Tracker Initiative). 2019a. *Balancing the Budget: Why Deflating the Carbon Bubble Requires Oil and Gas Companies to Shrink.* London: Carbon Tracker Initiative.

CTI. 2019b. *Breaking the Habit: Why None of the Large Oil Companies Are 'Paris-Aligned', and What They Need to Do to Get There.* London: Carbon Tracker Initiative.

CTI. 2020a. *Absolute Impact: Why Oil Majors' Climate Ambitions Fall Short of Paris Limits.* London: Carbon Tracker Initiative.

CTI. 2020b. *The Future's Not in Plastics: Why Plastics Demand Won't Rescue the Oil Sector.* London, Carbon Tracker Initiative.

CTI. 2021a. *Beyond Petrostates: The Burning Need to Cut Oil Dependence in the Energy Transition.* London: Carbon Tracker Initiative.

CTI. 2021b. *The Sky's the Limit: Solar and Wind Energy Potential Is 100 Times as Much as Global Energy Demand.* London: Carbon Tracker Initiative.

CTI. 2021c. *A Tale of Two Share Issues: How Fossil Fuel Equity Offerings Are Losing Investors Billions.* London: Carbon Tracker Initiative.

Cummings McLean, D. 2019. "Pope Francis Claims Climate in State of 'Emergency,' Asks World to 'Abandon' Fossil Fuels." *LifeSiteNews*, September 4, https://www

.lifesitenews.com/news/pope-francis-claims-climate-in-state-of-emergency-asks -world-to-abandon-fossil-fuels.

Cunningham, N. 2018. "*Oil Majors See Profits Spike, Exxon Lags Behind.*" Oilprice .com, May 1, https://oilprice.com/Energy/Energy-General/Oil-Majors-See-Profits -Spike-Exxon-Lags-Behind.html.

Cuomo, C. J. 2011. "Climate Change, Vulnerability, and Responsibility." *Hypatia* 26 (4): 690–714.

Cuvelier, L., and L. Pinson. 2021. *One Year On: BlackRock Still Addicted to Fossil Fuels. Reclaim Finance and Urgewald.* https://reclaimfinance.org/site/en/2021/01 /14/one-year-on-blackrock-still-addicted-to-fossil-fuels/.

Dafermos, Y., M. Nikolaidi, and G. Galanis. 2018. *Can Green Quantitative Easing (QE) Reduce Global Warming?* FEPS Policy Brief with GPERC.

D'Angelo, C. 2019. "Big Business Spent $1.4 Billion on PR, Advertising over the Last Decade." *Huffington Post,* March 14, https://www.huffpost.com/entry/industry -trade-groups-public-relations-spending_n_5c89aa69e4b0fbd76620a0a3.

Davenport, C., and J. Smialek. 2020. "Federal Report Warns of Financial Havoc from Climate Change." *New York Times,* September 8, https://www.nytimes.com /2020/09/08/climate/climate-change-financial-markets.html.

Davidson, M. D. 2008. "Parallels in Reactionary Argumentation in the US Congressional Debates on the Abolition of Slavery and the Kyoto Protocol." *Climatic Change* 86 (1–2): 67–82.

Davies, R. 2019. "Norway's $1tn Wealth Fund to Divest from Oil and Gas Exploration." *The Guardian,* March 8, https://www.theguardian.com/world/2019/mar /08/norways-1tn-wealth-fund-to-divest-from-oil-and-gas-exploration.

Deaton, J. 2017. "The Desperate but Effective Attempts to Silence Climate Scientists." *EcoWatch,* September 30, https://www.ecowatch.com/silence-climate -scientists-2491043851.html.

Deich, N., and D. Reali. 2019. "*Big Oil Is Funding Future Climate Tech—but Should They?*" Carbon 180, August 23, https://medium.com/@carbon180/big-oil-is-funding -future-climate-tech-but-should-they-c103aed36011.

Della Porta, D., and M. Diani. 2009. *Social Movements: An Introduction.* 2nd ed. Oxford, UK: Blackwell Publishing.

Demelle, B., and K. Grandia. 2016. "There Is No Doubt: Exxon Knew CO2 Pollution Was a Global Threat by Late 1970s." *DeSmog*, April 26, https://www.desmogblog.com/2016/04/26/there-no-doubt-exxon-knew-co2-pollution-was-global-threat-late-1970s.

Deng, S., S. Liu, X. Mo, L. Jiang, and P. Bauer-Gottwein. 2021. "Polar Drift in the 1990s Explained by Terrestrial Water Storage Changes." *Geophysical Research Letters*, e2020GL092114.

Diani, M. 1992. "The Concept of Social Movement." *Sociological Review* 40 (1): 1–25.

Dixon, G., O. Bullock, and D. Adams. 2019. "Unintended Effects of Emphasizing the Role of Climate Change in Recent Natural Disasters." *Environmental Communication* 13 (2): 135–143.

Dodd, L. 2019. "Religious Spearhead Campaign against Fossil Fuels." *The Tablet*, September 11, https://www.thetablet.co.uk/news/12061/religious-spearhead-campaign-against-fossil-fuels-.

Doherty, B. 2019. "Australia Fires Could Be Out of Control for Months, Says Fire Chief." *The Guardian*, November 12, https://www.theguardian.com/australia-news/2019/nov/12/australia-fires-rage-out-of-control-catastrophic-day.

Douglas, F. 1857. "West India Emancipation Speech, delivered at Canandaigua, New York, August 3, 1857." University of Rochester, Frederick Douglass Project. https://rbscp.lib.rochester.edu/4398.

Drugmand, D. 2019a. "Exxon Climate History on Trial: Oil Giant's Legal Challenges Reach Critical Mass This Fall." The Climate Docket, September 10, https://www.climatedocket.com/2019/09/10/exxon-climate-fraud-courts-lawsuits/.

Drugmand, D. 2019b. "Massachusetts Becomes Second State to Sue Exxon for Climate Fraud." The Climate Docket, October 24, https://www.climatedocket.com/2019/10/24/massachusetts-exxon-climate-fraud-maura-healey/.

Economist (The). 2020. "How Much Can Financiers Do about Climate Change?" *The Economist*, June 20, https://www.economist.com/briefing/2020/06/20/how-much-can-financiers-do-about-climate-change.

Economist (The) Intelligence Unit. 2021. *Democracy Index 2020: In Sickness and in Health?* https://www.eiu.com/n/campaigns/democracy-index-2020/.

Ekwurzel, B., J. Boneham, M. W. Dalton, R. Heede, R. J. Mera, M. R. Allen, and P. C. Frumhoff. 2017. "The Rise in Global Atmospheric CO2, Surface Temperature, and Sea Level from Emissions Traced to Major Carbon Producers." *Climatic Change* 144 (4): 579–590.

Elliott, L. 2019. "Tackling Climate Crisis Is What We Should Be Doing, Says New IMF Boss." *The Guardian*, November 30, https://www.theguardian.com /business/2019/nov/30/imf-boss-kristalina-georgiva-climate-crisis-financial -crash-economics?CMP=share_btn_tw.

Energy In Depth. 2021. "About EID." https://www.energyindepth.org/about/.

Erickson, P., M. Lazarus, and K. Tempest. 2015a. *Carbon Lock-In from Fossil Fuel Supply Infrastructure*. SEI Discussion Brief.

Erickson, P., S. Kartha, M. Lazarus, and K. Tempest. 2015b. "Assessing Carbon Lock-In." *Environmental Research Letters* 10 (8): 084023.

Espiner. 2019. *"Can Big Investors Save the World?"* BBC News-Business, August 18. https://www.bbc.com/news/business-49330150.

European Commission. 2020. *"The European Green Deal Investment Plan and Just Transition Mechanism Explained."* European Commission, January 14, https://ec .europa.eu/commission/presscorner/detail/en/qanda_20_24.

European Investment Bank. 2019. "EU Bank Launches Ambitious New Climate Strategy and Energy Lending Policy." November 14, https://www.eib.org/en/press /all/2019-313-eu-bank-launches-ambitious-new-climate-strategy-and-energy -lending-policy.

ExxonMobil. 2016. "Statement on Paris Climate Agreement Entering into Force." *Press release,* November 4, https://corporate.exxonmobil.com/Sustainability/Environmental -protection/Climate-change/Statements-on-Paris-climate-agreement.

ExxonMobil. 2019a. "ExxonMobil to Increase, Accelerate Permian Output to 1 Million Barrels per Day by 2024." *ExxonMobil Newsroom,* March 5, https://corporate .exxonmobil.com/news/newsroom/news-releases/2019/0305_exxonmobil-to -increase-accelerate-permian-output-to-1-million-barrels-per-day-by-2024.

ExxonMobil. 2019b. *"Outlook for Energy: A Perspective to 2040."* August 28, https:// corporate.exxonmobil.com/Energy-and-environment/Looking-forward/Outlook -for-Energy/Outlook-for-Energy-A-perspective-to-2040.

Fagan, M. 2000. "Sheikh Yamani Predicts Price Crash as Age of Oil Ends." *The Telegraph,* June 25, https://www.telegraph.co.uk/news/uknews/1344832/Sheikh -Yamani-predicts-price-crash-as-age-of-oil-ends.html.

Farand, C. 2018. "Oil and Money Conference: Fossil Fuel Executives Discuss How to Use PR to 'Combat Growing Divestment Movement.'" *DeSmog,* October 9, https://www.desmog.co.uk/2018/10/09/oil-and-money-conference-fossil-fuel -executives-discuss-how-use-pr-combat-growing-divestment-movement.

Farand, C. 2019. "European Central Bank Should 'Gradually Eliminate' Carbon Assets: Lagarde." *Climate Home News,* April 9, https://www.climatechangenews.com/2019 /09/04/european-central-bank-gradually-eliminate-carbon-assets-says-lagarde/.

Farand, C. 2020. "EIB Approves €1 Trillion Green Investment Plan to Become 'Climate Bank." *Climate Home News,* December 11, https://www.climatechangenews.com /2020/11/12/eib-approves-e1-trillion-green-investment-plan-become-climate-bank/.

Farmer, J. D., C. Hepburn, M. C. Ives, T. Hale, T. Wetzer, P. Mealy, and R. Way. 2019. "Sensitive Intervention Points in the Post-Carbon Transition." *Science* 364 (6436): 132–134.

Farrell, J. 2016. "Corporate Funding and Ideological Polarization about Climate Change." *Proceedings of the National Academy of Sciences* 113 (1): 92–97.

Feinberg, J. 1970. *Doing and Deserving.* Princeton, NJ: Princeton University Press.

Finnemore, M., and K. Sikkink. 1998. "International Norm Dynamics and Political Change." *International Organization* 52 (4): 887–917.

Fitzgibbon, W., M. Patrucić, and M. G. Rey. 2016. "How Family That Runs Azerbaijan Built an Empire of Hidden Wealth." International Consortium of Investigative Journalists, April 4, https://www.icij.org/investigations/panama-papers /20160404-azerbaijan-hidden-wealth/.

Foot, P. 1967. "The Problem of Abortion and the Doctrine of Double Effect." In *Virtues and Vices and Other Essays in Moral Philosophy,* ed. B. Steinbock and A. Norcross, 5–15. Oxford, UK: Blackwell.

Fortune. 2020. "Global 500." https://fortune.com/global500/.

Fossil Free. n.d. "1200+ Divestment Commitments." https://gofossilfree.org/divestment /commitments/. Accessed on September 25, 2021.

Franta, B. 2018. "Early Oil Industry Knowledge of CO2 and Global Warming." *Nature Climate Change* 8 (12): 1024–1025.

Franta, B. 2021. "Early Oil Industry Disinformation on Global Warming." *Environmental Politics* 30 (4): 663–668.

Fraser, N. 1995. "From Redistribution to Recognition: Dilemmas of Justice in a 'Post-Socialist' Age." *New Left Review* 212: 68–93.

Freedman, A. 2021. "Al Gore's Climate TRACE Finds Vast Undercounts of Emissions." *Axios Energy & Environment*, September 16, https://www.axios.com/globa -carbon-emissions-inventory-surprises-cb7f220a-6dfd-4f88-9349-5c9ffa0817e9.html.

Freeman, R. E. 1984. *Strategic Management: A Stakeholder Approach*. Marshfield MA: Pitman.

French, P. 1984. *Collective and Corporate Responsibility*. New York: Columbia University Press.

Friedman, M. 1962. *Capitalism and Freedom*. Chicago: University of Chicago Press

Friedman, M. 1970. "The Social Responsibility of Business Is to Increase Its Profits." *New York Times*, September 13.

Friends of the Earth, Bailout Watch, and Oxfam. 2021. *12 Guilty Fogeys: Big Oil's $86 Billion Offshore Tax Bonanza*. https://1bps6437gg8c169ioy1drtgz-wpengine .netdna-ssl.com/wp-content/uploads/2021/09/FFS_12_Guilty_Fogeys_rd3.pdf.

Frumhoff, P. C., and M. Allen. 2017. "Big Oil Must Pay for Climate Change. Now We Can Calculate How Much." *The Guardian*, September 7, https://www .theguardian.com/commentisfree/2017/sep/07/big-oil-must-pay-for-climate -change-here-is-how-to-calculate-how-much.

Frumhoff, P. C., R. Heede, and N. Oreskes. 2015. "The Climate Responsibilities of Industrial Carbon Producers." *Climatic Change* 132 (2): 157–171.

Gardiner, S. M. 2004. "Ethics and Global Climate Change." *Ethics* 114: 555–600.

Gaulin, N., and P. Le Billon. 2020. "Climate Change and Fossil Fuel Production Cuts: Assessing Global Supply-Side Constraints and Policy Implications." *Climate Policy* 20 (8): 888–901.

Geels, F. W. 2014. "Reconceptualising the Co-Evolution of Firms-in-Industries and Their Environments: Developing an Inter-Disciplinary Triple Embeddedness Framework." *Research Policy* 43 (2): 261–277.

Geels, F. W., F. Berkhout, and D. P. Vuuren. 2016. "Bridging Analytical Approaches for Low-Carbon Transitions." *Nature Climate Change* 6 (6): 576–583.

GGON (Global Gas and Oil Network). 2019. *Oil, Gas and the Climate: An Analysis of Oil and Gas Industry Plans for Expansion and Compatibility with Global Emission Limits.*

http://ggon.org/wp-content/uploads/2019/12/GGON19.OilGasClimate.EnglishFinal
.pdf.

Giblom, K. 2019. "Big Money Starts to Dump Stocks That Pose Climate Risks." *Bloomberg*, August 7, https://www.bloomberg.com/news/articles/2019-08-07/big -money-starts-to-dump-stocks-that-pose-climate-risks.

Gilding, P. 2019. "Why I Welcome a Climate Emergency." *Nature* 573: 311.

GlobalData. 2018. *Global Planned Oil and Gas Pipelines Industry Outlook to 2022—Capacity and Capital Expenditure Forecasts with Details of All Planned Pipe- lines.* London: GlobalData.

Global Witness. 2019. *Overexposed: The IPCC's Report on 1.5°C and the Risks of Overinvestment in Oil and Gas.* London: Global Witness.

Goldberg, M. H., J. R. Marlon, X. Wang, S. van der Linden, S., and A. Leiserowitz. 2020. "Oil and Gas Companies Invest in Legislators that Vote against the Envi- ronment." *Proceedings of the National Academy of Sciences* 117 (10): 5111–5112.

Goldthau, A., K. Westphal, M. Bazilian, and M. Bradshaw. 2019. "How the Energy Transition Will Reshape Geopolitics." *Nature* 569 (7754): 29–31.

Goodin, R. E. 1989. "Theories of Compensation." *Oxford Journal of Legal Studies* 9: 56–75.

Goodin, R. E. 2013. "Disgorging the Fruits of Historical Wrongdoing." *American Political Science Review* 107 (3): 478–491.

Gore, A. 2007. "Moving beyond Kyoto." *New York Times*, July 1, https://www .nytimes.com/2007/07/01/opinion/01gore.html.

Gramsci, A. 1975. *Quaderni dal Carcere.* Torino: Einaudi.

Grant, R. W., and R. O. Keohane. 2005. "Accountability and Abuses of Power in World Politics." *American Political Science Review* 99 (1): 29–43.

Grantham Research Institute. n.d. "Climate Change Laws of the World." Data- base, https://climate-laws.org/. Accessed on September 25, 2021.

Green, F. 2018a. "Anti-Fossil Fuel Norms." *Climatic Change* 150 (1–2): 103–116.

Green, F. 2018b. "The Logic of Fossil Fuel Bans." *Nature Climate Change* 8: 449–451.

Green, F., and E. Brandstedt. 2021. "Engaged Climate Ethics." *Journal of Political Philosophy* 29 (4): 539–563.

Green, F., and R. Dennis. 2018. "Cutting with Both Arms of the Scissors: The Economic and Political Case for Restrictive Supply-Side Climate Policies." *Climatic Change* 150 (1–2): 73–87.

Green, J. F., J. Hadden, T. Hale, and P. Mahdavi. 2020. "Transition, Hedge, or Resist? Understanding Political and Economic Behavior toward Decarbonization in the Oil and Gas Industry." SSRN.com, September 17, https://papers.ssrn.com /sol3/papers.cfm?abstract_id=3694447.

Green, M. 2019. "130 Banks Worth $47 Trillion Adopt New UN-Backed Climate Policies to Shift Their Loan Books Away from Fossil Fuels." *Reuters—Business Insider*, September 22, https://www.businessinsider.com/banks -worth-47-trillion-adopt-new-un-backed-climate-principles-2019-9?IR=T &sfns=mo.

Greenfield, P. 2019. "Fossil Fuel Bosses Must Change or Be Voted Out, Says Asset Manage." *The Guardian*, October 12, https://www.theguardian.com/environmen /2019/oct/12/fossil-fuel-bosses-change-voted-out-asset-manager.

Guarascio, F. 2019. "Worried by Climate Change, EU Moves to End Fossil Fuel Funding." *Reuters World News,* November 8, https://uk.reuters.com/article/uk -climatechange-europe-eib/worried-by-climate-change-eu-moves-to-end-fossil -fuel-funding-idUKKBN1XI1Z0.

Gunningham, N. 2017. "Building Norms from the Grassroots Up: Divestment, Expressive Politics, and Climate Change." *Law & Policy* 39 (4): 372–392.

Gunningham, N., R. A. Kagan, and D. Thornton. 2004. "Social License and Environmental Protection: Why Businesses Go beyond Compliance." *Law & Social Inquiry* 29 (2): 307–341.

Guthrie, J. 2020. "Lex in Depth: Examining the Slave Trade—'Britain Has a Debt to Repay." *Financial Times*, June 27, https://www.ft.com/content/945c6136-0b93 -41bf-bd80-a80d944bb0b8.

Hacker, J. D., and J. M. McPherson. 2011. "A Census-Based Count of the Civil War Dead: With Introductory Remarks by James M. McPherson." *Civil War History* 57 (4): 307–348.

Hadden, J. 2014. "Explaining Variation in Transnational Climate Change Activism: The Role of Inter-Movement Spillover." *Global Environmental Politics* 14: 7–25.

Hanna, R., Y. Xu, and D. G. Victor. 2020. "After COVID-19, Green Investment Must Deliver Jobs to Get Political Traction." *Nature* 582: 178–180.

Hansen, J. 2011. "Obama's Second Chance on the Predominant Moral Issue of This Century." Huffington Post, May 25, https://www.huffpost.com/entry/obamas-second-chance-on-c_b_525567.

Hansson, S. O. 2018. "Dealing with Climate Science Denialism: Experiences from Confrontations with Other Forms of Pseudoscience." *Climate Policy* 18 (9): 1094–1102.

Harder, A. 2020. "BP's Climate Reinvention Dodges Politics." *Axios Energy & Environment*, October 13, https://www.axios.com/bp-ceo-climate-reinvention-axios-on-1b0-624c3a9c-9ad4-4fb1-9b38-552c558a3544.html.

Harding, L. 2007. "Putin, the Kremlin Power Struggle and the $40bn Fortune." *The Guardian*, December 20, https://www.theguardian.com/world/2007/dec/21/russia.topstories3.

Harper, W. 1838. *Memoir on Slavery: Read before the Society for the Advancement of Learning, of South Carolina, at Its Annual Meeting at Columbia, 1837.* Charleston, SC: James S. Burges.

Hart, H. L. A. 1963. *Law, Liberty, and Morality.* Stanford, CA: Stanford University Press.

Hasemyer, D. 2020. "Minnesota and the District of Columbia Allege Climate Change Deception by Big Oil." *Inside Climate News*, June 25, https://insideclimatenews.org/news/24062020/minnesota-climate-change-lawsuit-exxon-mobil-api-koch-industries.

Hasemyer, D. 2021. "Maryland's Capital City Joins a Long Line of Litigants Seeking Climate-Related Damages from the Fossil Fuel Industry." *Inside Climate News*, February 24, https://insideclimatenews.org/news/24022021/annapolis-maryland-climate-oil-lawsuit/.

Hayes, C. 2014. "The New Abolitionism." *The Nation*, April 22, https://www.thenation.com/article/new-abolitionism/?print=1.

Heal, G., and W. Schlenker. 2019. *"Coase, Hotelling and Pigou: The Incidence of a Carbon Tax and CO2 Emissions."* Working paper 26086. https://www.nber.org/papers/w26086.pdf.

Healy, N., and J. Barry. 2017. "Politicizing Energy Justice and Energy System Transitions: Fossil Fuel Divestment and a 'Just Transition.'" *Energy Policy* 108: 451–459.

Heede, R. 2013. *Carbon Majors: Accounting for Carbon and Methane Emissions 1854-2010.* Snowmass, CO: Climate Mitigation Services. https://www.greenpeace.org/static/planet4-philippines-stateless/2019/05/2ae27cd6-mrr-8.3-7nov13.pdf.

Heede, R. 2014. "Tracing Anthropogenic Carbon Dioxide and Methane Emissions to Fossil Fuel and Cement Producers, 1854–2010." *Climatic Change* 122 (1–2): 229–241.

Heffron, R. J., and D. McCauley. 2018. "What Is the 'Just Transition'?" *Geoforum* 88: 74–77.

Herring, S. C., N. Christidis, A. Hoell, M. P. Hoerling, and P. A. Stott. 2021 "Explaining Extreme Events of 2019 from a Climate Perspective." *Bulletin of the American Meteorological Society* 102 (1): 1–116.

Hess, D. J. 2014. "Sustainability Transitions: A Political Coalition Perspective.' *Research Policy* 43 (2): 278–283.

Hmiel, B., V. V. Petrenko, M. N. Dyonisius, C. Buizert, A. M. Smith, P. F. Place, C Harth, R. Beaudette, Q. Hua, B. Yang, I. Vimont, S. E. Michel, J. P. Severing haus, D. Etheridge, T. Bromley, J. Schmitt, X. Faïn, R. F. Weiss, and E. Dlugo kencky. 2020. "Preindustrial 14CH4 Indicates Greater Anthropogenic Fossil CH_4 Emissions." *Nature* 578: 409–412.

Hoffarth, M. R., and G. Hodson. 2016. "Green on the Outside, Red on the inside Perceived Environmentalist Threat as a Factor Explaining Political Polarization o¯ Climate Change." *Journal of Environmental Psychology* 45: 40–49.

Hope, M. 2018. "What #ShellKnew and How It Was Used to Stall International Climate Change Negotiations." *DeSmog*, July 10, https://www.desmogblog.com /2018/07/10/what-shellknew-and-how-it-was-used-stall-international-climate -change-negotiations.

Hope, M. 2019a. "How Big Oil Tried to Capture the UN Intergovernmental Pane on Climate Change." *DeSmog*, April 24, https://www.desmogblog.com/2019/0< /24/how-big-oil-tried-failed-capture-un-intergovernmental-panel-climate-change

Hope, M. 2019b. "Polly Higgins—Meet the Lawyer Taking on Big Oil's 'Crime: Against Humanity.'" *DeSmog*, April 16, https://www.desmogblog.com/2019/0< /16/polly-higgins-lawyer-taking-big-oil-s-crimes-against-humanity.

Hormio, S. 2017. "Can Corporations Have (Moral) Responsibility regarding Cli mate Change Mitigation?" *Ethics, Policy & Environment* 20 (3): 314–332.

HRC (Human Rights Council). 2019. *"Climate Change and Poverty:* Special Rap porteur on Extreme Poverty and Human Rights." Extreme Poverty and Human Rights, June 19, https://srpovertyorg.files.wordpress.com/2019/06/unsr-povert¯ -climate-change-a_hrc_41_39.pdf.

Hsu, A., N. Höhne, T. Kuramochi, M. Roelfsema, A. Weinfurter, Y. Xie, and P. Faria. 2019. "A Research Roadmap for Quantifying Non-State and Subnational Climate Mitigation Action." *Nature Climate Change* 9 (1): 11.

Huggel, C., I. Wallimann-Helmer, D. Stone, and W. Cramer. 2016. "Reconciling Justice and Attribution Research to Advance Climate Policy." *Nature Climate Change* 6 (10): 901–908.

Hulme, M. 2011. "Reducing the Future to Climate: A Story of Climate Determinism and Reductionism." *Osiris* 26 (1): 245–266.

IBISWorld. 2020. "Global Oil & Gas Exploration & Production: *Global Market Research Report*." https://www.ibisworld.com/global/market-research-reports/global-oil-gas-exploration-production-industry/.

IEA (International Energy Agency). 2018. *The Future of Petrochemicals. Towards more Sustainable Plastics and Fertilisers*. Paris: IEA.

IEA. 2019a. *Global Energy & CO2 Status Report 2018*. Paris: IEA.

IEA. 2019b. *World Energy Investment 2019*. Paris: IEA.

IEA. 2019c. *World Energy Outlook 2019*. Paris: IEA.

IEA. 2020a. *Global Energy Review 2020: The Impacts of the Covid-19 Crisis on Global Energy Demand and CO2 Emissions*. Paris: IEA.

IEA. 2020b. *The Oil and Gas Industry in Energy Transitions*. Paris: IEA.

IEA. 2020c. *Renewables 2020: Analysis and Forecast to 2025*. Paris: IEA.

IEA. 2020d. "Report Extract: Sustainable Development Scenario." https://www.iea.org/reports/world-energy-model/sustainable-development-scenario.

IEA. 2020e. *World Energy Outlook 2020*. Paris: IEA.

IEA. 2020f. "World Energy Outlook 2020 Shows How the Response to the Covid Crisis Can Reshape the Future of Energy." Press release, October 13, https://www.iea.org/news/world-energy-outlook-2020-shows-how-the-response-to-the-covid-crisis-can-reshape-the-future-of-energy.

IEA. 2021a. *Global Energy Review: CO$_2$ Emissions in 2020*. Paris: IEA.

IEA. 2021b. *Global Energy Review 2021*. Paris: IEA.

IEA. 2021c. *Net Zero by 2050. A Roadmap for the Global Energy Sector*. Paris: IEA.

IEA. 2021d. *Oil. Analysis and Forecast to 2026*. Paris: IEA.

IFRC (International Federation of Red Cross and Red Crescent Societies). 2020. *World Disaster Report 2020: Come Heat or High Water*. Geneva: IFRC. https://media.ifrc.org/ifrc/world-disaster-report-2020/.

InfluenceMap. 2019. "Big Oil's Real Agenda on Climate Change. How the Oil Majors Have Spent $1Bn after Paris in Narrative Capture and Lobbying." March. https://influencemap.org/report/How-Big-Oil-Continues-to-Oppose-the-Paris-Agreement-38212275958aa21196dae3b76220bddc.

IMF (International Monetary Fund). 2021. "GDP Per Capita, Current Prices." https://www.imf.org/external/datamapper/NGDPDPC@WEO/OEMDC/ADVEC/WEOWORLD.

IPCC (International Panel on Climate Change). 1990. *Climate Change: The IPCC Scientific Assessment*. Geneva: IPCC.

IPCC. 1992. *Climate Change: The IPCC 1990 and 1992 Assessments*. Geneva: IPCC.

IPCC. 1995. *IPCC Second Assessment: Climate Change 1995*. Geneva: IPCC.

IPCC. 2014. *Working Group II, Impacts, Adaptation, and Vulnerability*. Cambridge: Cambridge University Press.

IPCC. 2018. *IPCC Special Report on Global Warming of 1.5°C*. Geneva: IPCC.

IPCC. 2021a. *Climate Change 2021: The Physical Science Basis. Contribution of Working Group I to the Sixth Assessment Report of the Intergovernmental Panel on Climate Change*. Geneva: IPCC.

IPCC. 2021b. *Summary for Policymakers. In: Climate Change 2021: The Physical Science Basis. Contribution of Working Group I to the Sixth Assessment Report of the Intergovernmental Panel on Climate Change*. Geneva: IPCC.

IPIECA (International Petroleum Industry Environmental Conservation Association). 2013. *Addressing Adaptation in the Oil and Gas Industry*. London: IPIECA.

IWG (Interagency Working Group on Social Cost of Greenhouse Gases). 2021. "Technical Support Document: Social Cost of Carbon, Methane, and Nitrous Oxide; Interim Estimates under Executive Order 13990." White House, February, https://www.whitehouse.gov/wp-content/uploads/2021/02/TechnicalSupportDocument_SocialCostofCarbonMethaneNitrousOxide.pdf?utm_source=newsletter&utm_medium=email&utm_campaign=newsletter_axiosgenerate&stream=top.

Jackson, R. B., P. Friedlingstein, R. M. Andrew, J. G. Canadell, C. Le Quéré, and G. P. Peters. (2019). "Persistent Fossil Fuel Growth Threatens the Paris Agreement and Planetary Health." *Environmental Research Letters* 14 (12): 121001.

Jaffe, A. M. 2020. "Striking Oil Ain't What It Used to Be. Poor Countries Find Fossil Fuels Just as the Rich World Swears Them Off." *Foreign Affairs* 99: 1. https://www.foreignaffairs.com/articles/africa/2020-01-20/striking-oil-aint-what-it-used-be.

James, R. A., R. G. Jones, E. Boyd, H. R. Young, F. E. Otto, C. Huggel, and J. S. Fuglestvedt. 2019. "Attribution: How Is It Relevant for Loss and Damage Policy and Practice?" In *Loss and Damage from Climate Change: Concepts, Methods and Policy Options,* ed. R. Mechler, L. M. Bouwer, T. Schinko, S. Surminski, and J. Linnerooth-Bayer, 113–154. Cham: Springer.

Jamieson, D. 2008. "The Post-Kyoto Climate: A Gloomy Forecast." *Georgetown International Environmental Law Review* 20: 537–551.

Jamieson, D. 2015. "Responsibility and Climate Change." *Global Justice: Theory Practice Rhetoric* 8 (2): 23–42.

Jamieson, D. 2017. "Slavery, Carbon, and Moral Progress." *Ethical Theory and Moral Practice* 20 (1): 169–183.

Jerving, S. 2019. "African Development Bank Commits to Coal-Free Financing." *Devex,* September 26, https://www.devex.com/news/african-development-bank-commits-to-coal-free-financing-95698.

Jerving, S., K. Jenning, M. M. Hirsch, and S. Rust. 2015. "What Exxon Knew about the Earth's Melting Arctic." *Los Angeles Times*, October 9, http://graphics.latimes.com/exxon-arctic/.

Juhasz, A. 2021. "Exxon's Oil Drilling Gamble Off Guyana Coast 'Poses Major Environmental Risk.'" *The Guardian*, August 17, https://www.theguardian.com/environment/2021/aug/17/exxon-oil-drilling-guyana-disaster-risk.

Kartha, S., S. Caney, N. K. Dubash, and G. Muttitt. 2018. "Whose Carbon Is Burnable? Equity Considerations in the Allocation of a 'Right to Extract.'" *Climatic Change* 150 (1–2): 117–129.

Keck, M. E., and K. Sikkink. 2014. *Activists beyond Borders: Advocacy Networks in International Politics*. Ithaca, NY: Cornell University Press.

Kelly, S. 2020. *After a Decade of Fracking, Billions of Dollars Lost and a Climate in Crisis*. DeSmog, January 21, https://www.desmog.co.uk/2020/01/21/2010s-decade-fracking-shale-climate-crisis.

Kenner, D. 2019. "The Polluter Elite Database." Why Green Economy?, June, http://whygreeneconomy.org/the-polluter-elite-database/.

Kirka, D. 2020. *"UN Envoy Lays out Strategy for Financing Climate Battle."* Associated Press, November 9, https://apnews.com/article/climate-climate-change -paris-mark-carney-united-nations-8b08cee6f3351d03ebd3130b9cb7a608.

Knutson, J. 2020. "Pope Calls on World to Act with 'Urgency' to Curb Climate Change." *Axios*, October 10, https://www.axios.com/pope-francis-climate-change -carbon-emissions-453d6e9a-9b73-419b-ae8a-476bd648ea15.html.

Knutson, T. 2017. *"Detection and Attribution Methodologies Overview."* In *Climate Science Special Report: Fourth National Climate Assessment, Volume I*, ed. D. J. Wuebbles, D. W. Fahey, K. A. Hibbard, D. J. Dokken, B. C. Stewart, and T. K. Maycock, 443–451. Washington, DC: US Global Change Research Program.

Köhler, J., F. W. Geels, F. Kern, J. Markard, E. Onsongo, A. Wieczorek, and P. Wells. 2019. "An Agenda for Sustainability Transitions Research: State of the Art and Future Directions." *Environmental Innovation and Societal Transitions* 31: 1–32.

Kolbert, E. 2021. "Biden's Jobs Plan Is Also a Climate Plan. Will It Make a Difference?" *New Yorker*, April 4.

Kornek, U., C. Kardish, C. Flachsland, S. Levi, and O. Edenhofer. 2020. "What Is Important for Achieving 2°C? UNFCCC and IPCC Expert Perceptions on Obstacles and Response Options for Climate Change Mitigation." *Environmental Research Letters* 15 (2): 024005.

Kramer, R. J., H. He, B. J. Soden, L. Oreopoulos, G. Myhre, P. M. Forster, and C. J. Smith. 2021. "Observational Evidence of Increasing Global Radiative Forcing." *Geophysical Research Letters* 48 (7): 091585.

Krauss, C. 2019. "Flood Oil Is Coming, Complicating Efforts to Fight Global Warming." *New York Times*, November 3, https://www.nytimes.com/2019/11/03 /business/energy-environment/oil-supply.html.

Kusnetz, N. 2020. "What Does Net Zero Emissions Mean for Big Oil? Not What You'd Think." *Inside Climate News*, July 16, https://insideclimatenews.org/news /15072020/oil-gas-climate-pledges-bp-shell-exxon.

Lamb, W., G. Mattioli, S. Levi, J. T. Roberts, S. Capstick, F. Creutzig, J. Minx, F. Müller-Hansen, T. Culhane, and J. Steinberger. 2020. "Discourses of Climate Delay." *Global Sustainability* 3: E17.

Landrum, L., and M. M. Holland. 2020. "Extremes Become Routine in an Emerging New Arctic." *Nature Climate Change* 10 (12): 1108–1115.

Lavallee, J. P., B. Di Giusto, and T. Y. Yu. 2019. "Collective Responsibility Framing Also Leads to Mitigation Behavior in East Asia: A Replication Study in Taiwan." *Climatic Change* 153: 423–438.

Lazarus, M., and H. van Asselt. 2018. "Fossil Fuel Supply and Climate Policy: Exploring the Road Less Taken." *Climatic Change* 150 (1–2): 1–13.

Leber, R. 2019. "*How a Revolution in Climate Science Is Putting Big Oil Back on Trial.*" *Mother Jones*, September 16, https://www.motherjones.com/politics/2019/09/how-a-revolution-in-climate-science-is-putting-big-oil-back-on-trial/.

Le Billon, P., and B. Kristoffersen. 2020. "Just Cuts for Fossil Fuels? Supply-Side Carbon Constraints and Energy Transition." *Environment and Planning A* 52 (6): 1072–1092.

Lenferna, G. A. 2018. "Can We Equitably Manage the End of the Fossil Fuel Era?" *Energy Research & Social Science* 35: 217–223.

Le Quéré, C., R. M. Andrew, P. Friedlingstein, S. Sitch, J. Pongratz, A. C. Manning, and T. A. Boden. 2018. "Global Carbon Budget 2017." *Earth System Science Data* 10: 405–448.

Le Quéré, C., R. B. Jackson, and M. W. Jones. 2020a. "Temporary Reduction in Daily Global CO_2 Emissions during the COVID-19 Forced Confinement." *Nature Climate Change* 10: 647–653.

Le Quéré, C., R. B. Jackson, M. W. Jones, A. J. P. Smith, S. Abernethy, R. M. Andrew, A. J. De-Gol, et al. 2020b. *Supplementary Data to: Le Quéré et al.* https://doi:10.18160/RQDW-BTJU.

Le Quéré, C., G. P. Peters, P. Friedlingstein, R. M. Andrew, J. G. Canadell, S. J. Davis, and M. W. Jones. 2021. "Fossil CO_2 Emissions in the Post-COVID-19 Era." *Nature Climate Change* 11 (3): 197–199.

Levy, D. L., and D. Egan. 2003. "A Neo-Gramscian Approach to Corporate Political Strategy: Conflict and Accommodation in the Climate Change Negotiations." *Journal of Management Studies* 40 (4): 803–829.

Levy, D. L., and P. J. Newell. 2002. "Business Strategy and International Environmental Governance: Toward a Neo-Gramscian Synthesis." *Global Environmental Politics* 2 (4): 84–101.

Levy, D. L., and P. J. Newell. 2005. "A Neo-Gramscian Approach to Business in International Environmental Politics: An Interdisciplinary, Multilevel Framework." In *The Business of Global Environmental Governance*, ed. D. L. Levy and P. J. Newell, 47–72. Cambridge, MA: MIT Press.

Ley, A. J. 2018. "Mobilizing Doubt: The Legal Mobilization of Climate Denialist Groups." *Law & Policy* 40 (3): 221–242.

Leylim, C. 2021. "A New Hope? Launch of the Beyond Oil and Gas Alliance." 350.org, September 16, https://350.org/a-new-hope-launch-of-the-beyond-oil-and-gas-alliance/.

Licker, R., B. Ekwurzel, S. C. Doney, S. R. Cooley, I. D. Lima, R. Heede, and P. C. Frumhoff. 2019. "Attributing Ocean Acidification to Major Carbon Producers." *Environmental Research Letters* 14 (12): 124060.

Lindblom, C. E. 2001. *The Market System: What It Is, How It Works, and What to Make of It*. New Haven, CT: Yale University Press.

Lo, J. 2021. "Star Wars without Darth Vader'—Why the UN Climate Science Story Names No Villains." *Climate Home News*, December 1, https://www.climatechangenews.com/2021/01/12/star-wars-without-darth-vader-un-climate-science-story-names-no-villains/.

Lofoten Declaration (The). 2017. "The Lofoten Declaration: Climate Leadership Requires a Managed Decline of Fossil Fuel Production." http://www.lofotendeclaration.org/#read.

Lovell, B. 2010. *Challenged by Carbon: The Oil Industry and Climate Change*. Cambridge: Cambridge University Press.

MacDonald, C. 2020. "Exxon Spends Millions on Facebook to Keep the Fossil Fuel Industry Alive." *These Times*, October 20, https://inthesetimes.com/article/exxon-facebook-instagram-advertising-fracking-climate-fossil-fuels.

Mace, M. 2019. "Oil and Gas Companies Risking £1.8trn in Stranded Assets during Low-Carbon Transition, Report Warns." *Edie Newsroom*, September 6, https://www.edie.net/news/6/Oil-and-gas-companies-risking--2-2trn-in-stranded-assets-during-low-carbon-transition--report-warns/.

MacNeil, R., and M. Paterson. 2020. "Trump, US Climate Politics, and the Evolving Pattern of Global Climate Governance." *Global Change, Peace & Security* 32 (1): 1–18.

Manley, D., and P. R. P. Heller. 2021. "Risky Bet: National Oil Companies in the Energy Transition." *Natural Resource Governance Institute*, February 9, https://resou

rngovernance.org/analysis-tools/publications/risky-bet-national-oil-companies
-energy-transition.

Mann, M. E. 2019. "Lifestyle Changes Aren't Enough to Save the Planet. Here's
What Could." *Time*, September 12, https://time.com/5669071/lifestyle-changes
-climate-change/.

Marjanac, S., L. Patton, and J. Thornton. 2017. "Acts of God, Human Influence
and Litigation." *Nature Geoscience* 10 (9): 616.

Mark, J. 2018. "The Case for Climate Reparations: Who Should Pay the Costs
for Climate-Change–Related Disasters." *Sierra Magazine*, April 23, https://www
.sierraclub.org/sierra/2018-3-may-june/feature/the-case-for-climate-reparations.

McAdam, D. 2017. "Social Movement Theory and the Prospects for Climate Change
Activism in the United States." *Annual Review of Political Science* 20: 189–208.

McDuff, P. 2018. "Climate Change Denial Won't Even Benefit Oil Companies
Soon." *The Guardian*, July 31, https://www.theguardian.com/commentisfree/2018
/jul/31/climate-change-denial-oil-companies-fossil-fuels.

McKibben, B. 2016. "Let's Give up the Climate Change Charade: Exxon Won't Change
Its Stripes." *The Guardian*, May 20, https://www.theguardian.com/commentisfree
/2016/may/20/exxon-shareholders-climate-change-reform-divest.

McKibben, B. 2019. "Money Is the Oxygen on Which the Fire of Global Warming Burns." *New Yorker*, September 17, https://www.newyorker.com/news/daily
-comment/money-is-the-oxygen-on-which-the-fire-of-global-warming-burns.

McKibben, B. 2020a. *"Are We Past the Peak of Big Oil's Power?" New Yorker*, May
28, https://www.newyorker.com/news/annals-of-a-warming-planet/are-we-past
-the-peak-of-big-oils-power.

McKibben, B. 2020b. *"The Climate Crisis." New Yorker*, February 27, https://www
.newyorker.com/news/annals-of-a-warming-planet/welcome-to-the-climate-crisis
-newsletter.

Mechler, R., and T. Schinko. 2016. "Identifying the Policy Space for Climate Loss
and Damage." *Science* 354 (6310): 290–292.

Mecklin, J., ed. 2021. "This Is Your COVID Wake-up Call: It Is 100 Seconds to
Midnight; 2021 Doomsday Clock Statement." https://thebulletin.org/wp-content
/uploads/2021/01/2021-doomsday-clock-statement-1.pdf.

Meckling, J. 2015. "Oppose, Support, or Hedge? Distributional Effects, Regulatory Pressure, and Business Strategy in Environmental Politics." *Global Environmental Politics* 15 (2): 19–37.

Meckling, J. 2019. "A New Path for US Climate Politics: Choosing Policies That Mobilize Business for Decarbonization." *ANNALS of the American Academy of Political and Social Science* 685 (1): 82–95.

Mehta, J., and J. Jackson. 2021. "To Stop Climate Disaster, Make Ecocide an International Crime. It's the Only Way." *The Guardian*, February 24, https://www.theguardian.com/commentisfree/2021/feb/24/climate-crisis-ecocide-international-crime.

Meredith, S. 2019. "Our Biggest Compliment Yet': Greta Thunberg Thanks OPEC for Criticism." *CNBC*, March 5, https://www.cnbc.com/2019/07/05/greta-thunberg-thanks-oil-cartel-opec-for-climate-change-criticism.html?__source=sharebar.

Metcalf, G. E., and D. Weisbach. 2009. "The Design of a Carbon Tax." *Harvard Environmental Law Review* 33: 499–556.

Mikulka, J. 2020. "Major Fossil Fuel PR Group is Behind Europe Pro-Hydrogen Push." DeSmog, December 9, https://www.desmog.com/2020/12/09/fti-consulting-fossil-fuel-pr-group-behind-europe-hydrogen-lobby/.

Millar, R. J., C. Hepburn, J. Beddington, and M. R. Allen. 2018. "Principles to Guide Investment towards a Stable Climate." *Nature Climate Change* 8 (1): 2–4.

Millard, P. 2020. "*Petrobras CEO Calls Net Zero a Fad, Echoing Exxon's Focus on Oil.*" Bloomberg Green, December 2, https://www.bloomberg.com/news/articles/2020-12-02/petrobras-ceo-calls-net-zero-a-fad-echoing-exxon-s-focus-on-oil.

Miller, D. 2007. *National Responsibility and Global Justice.* Oxford: Oxford University Press.

Miller, D. 2008. "*Global Justice and Climate Change: How Should Responsibilities Be Distributed.*" *Tanner Lecture on Human Values* 28: 119–156, https://tannerlectures.utah.edu/_resources/documents/a-to-z/m/Miller_08.pdf.

Milman, O. 2021. "Oil Firms Knew Decades Ago Fossil Fuels Posed Grave Health Risks, Files Reveal." *The Guardian*, March 18, https://www.theguardian.com/environment/2021/mar/18/oil-industry-fossil-fuels-air-pollution-documents.

Mitchell, T. 2013. *Carbon Democracy: Political Power in the Age of Oil.* London: Verso.

Moody's. 2020. "Moody's—Energy Transition Poses Varying Degrees of Credit Risk to National Oil Companies." *Moody's Investor Service*, October 5, https://www.moodys.com/research/Moodys-Energy-transition-poses-varying-degrees-of-credit-risk-to-PBC_1248664.

Morgan, C. 2020. "Radical Object: Covert Broadcasts and the Nuclear Disarmament Campaign." History Workshop, January 22, https://www.historyworkshop.org.uk/radical-objects-covert-broadcasts-and-the-nuclear-disarmament-campaign/.

Mouhot, J. F. 2011. "Past Connections and Present Similarities in Slave Ownership and Fossil Fuel Usage." *Climatic Change* 105 (1–2): 329–355.

Mower, J., and T. Bradner. 2020. "*Shell to Resume Oil and Gas Exploration in Alaska Arctic Offshore.*" S&P Global Platts, September 17, https://www.spglobal.com/platts/en/market-insights/latest-news/natural-gas/091720-shell-to-resume-oil-and-gas-exploration-in-alaska-arctic-offshore.

Mulvey, K., M. Allen, and P. C. Frumhoff. 2019. "Fossil Fuel Companies Claim They're Helping Fight Climate Change. The Reality Is Different." *Bulletin of the Atomic Scientists*, December 17, https://thebulletin.org/2019/12/fossil-fuel-companies-claim-theyre-helping-fight-climate-change-the-reality-is-different/#.

Mulvey, K., J. Piepenburg, G. Goldman, and P. C. Frumhoff. 2016. "*The Climate Accountability Scorecard: Ranking Major Fossil Fuel Companies on Climate Deception, Disclosure, and Action.*" Union of Concerned Scientists, October 3, https://www.ucsusa.org/global-warming/fight-misinformation/climate-accountability-scorecard-ranking-major-fossil-fuel-companies#.WzIKSKl9jdc.

Muttitt, G. 2016. "*The Sky's Limit. Why the Paris Climate Goals Require a Managed Decline of Fossil Fuel Production.*" Oil Change International, September 22, http://priceofoil.org/2016/09/22/the-skys-limit-report/.

Myers, K. F., P. T. Doran, J. Cook, J. E. Kotcher, and T. A. Myers. 2021. "Consensus Revisited: Quantifying Scientific Agreement on Climate Change and Climate Expertise among Earth Scientists 10 Years Later." *Environmental Research Letters* 16 (10): 104030.

NAS (National Academies of Sciences, Engineering, and Medicine). 2016. *Attribution of Extreme Weather Events in the Context of Climate Change.* Washington, DC: National Academies Press. https://doi:10.17226/21852.

Nasiritousi, N. 2017. "Fossil Fuel Emitters and Climate Change: Unpacking the Governance Activities of Large Oil and Gas Companies." *Environmental Politics* 26 (4): 621–647.

National Intelligence Council. 2021. *Global Trends 2040: A More Contested World*. Washington DC, NIC, https://www.dni.gov/files/ODNI/documents/assessments /GlobalTrends_2040.pdf.

Natural Resource Governance Institute. 2021. "National Oil Company Database." https://www.nationaloilcompanydata.org/.

Newell, P., and D. Mulvaney. 2013. "The Political Economy of the 'Just Transition.'" *Geographical Journal* 179 (2): 132–140.

Newell, P., and M. Paterson. 1998. "A Climate for Business: Global Warming, the State and Capital." *Review of International Political Economy* 5 (4): 679–703.

Newell, P., and A. Simms. 2020. "Towards a Fossil Fuel Non-Proliferation Treaty." *Climate Policy* 20 (8): 1043–1054.

New York City Government. 2018. "Transcript: Mayor de Blasio Announces First-in-the-Nation Goal to Divest from Fossil Fuels." January 10, https://www1.nyc.gov /office-of-the-mayor/news/024-18/transcript-mayor-de-blasio-first-in-the-nation -goal-divest-fossil-fuels.

NOAA (National Oceanic and Atmospheric Administration, National Centers for Environmental Information). 2021. "U.S. Billion-Dollar Weather and Climate Disasters." Washington DC, NOAA.

Normile, D. 2021. "China Announces Major Boost for R&D, but Plan Lacks Ambitious Climate Targets." *Science*. https://doi:10.1126/science.abh3988.

Nuccitelli, D. 2015. "Two-Faced Exxon: The Misinformation Campaign against Its Own Scientist." *The Guardian*, November 25, https://www.theguardian.com /environment/climate-consensus-97-per-cent/2015/nov/25/two-faced-exxon-the -misinformation-campaign-against-its-own-scientists.

Obradovich, N., and S. M. Guenther. 2016. "Collective Responsibility Amplifies Mitigation Behaviors." *Climatic Change* 137 (1–2): 307–319.

OCI (Oil Change International). 2019. "Response to International Energy Agency's 2019 World Energy Outlook." Press release, November 13, http://priceofoil .org/2019/11/13/iea-weo-response/.

OCI. 2020. *Big Oil Reality Check: Assessing Oil and Gas Climate Plans*. Washington, DC: Oil Change International.

Ogen, Y. 2020. "Assessing Nitrogen Dioxide (NO_2) Levels as a Contributing Factor to the Coronavirus (COVID-19) Fatality Rate." *Science of the Total Environment* 138605.

Oil & Gas Journal. 2020a. "2020 Survey of Top 100 Non-US Oil & Gas Companies." September 7, https://www.ogj.com/ogj-survey-downloads/ogj-top-100-non-us-oil-gas-companies/document/14183408/2020-survey-of-top-100-nonus-oil-gas-companies.

Oil & Gas Journal. 2020b. "2020 Survey of Top 150 US Oil & Gas Companies." September 2, https://www.ogj.com/ogj-survey-downloads/ogj-top-200-us-oil-gas-companies/document/14074708/2019-survey-of-top-150-us-oil-gas-companies.

Oil & Gas UK. 2020. "Decommissioning Insights 2020." https://oguk.org.uk/wp-content/uploads/2020/11/Decommissioning-Insight.pdf.

O'Neill, O. 2001. "Agents of Justice." Metaphilosophy 32 (1–2): 180–195.

O'Neill, O. 2005. "The Dark Side of Human Rights." International Affairs 81 (2): 427–439.

OPEC (Organization of the Petroleum Exporting Countries). 2020. 2020 World Oil Outlook 2045. https://www.opec.org/opec_web/static_files_project/media/downloads/press_room/Launch%20of%20the%20WOO2020%20-%20presentation.pdf.

Oreskes, N. 2019. "Opinion: Can We Please Base Our Climate Change Discussions on Facts?" Los Angeles Times, September 12, https://www.latimes.com/opinion/story/2019-09-11/climate-change-false-assumptions-nuclear-power-fossil-fuels.

Oreskes, N., and E. M. Conway. 2011. Merchants of Doubt: How a Handful of Scientists Obscured the Truth on Issues from Tobacco Smoke to Global Warming. New York: Bloomsbury Publishing.

Oxfam 2020. Confronting Carbon Inequality. Nairobi: Oxfam International. https://oxfamilibrary.openrepository.com/bitstream/handle/10546/621052/mb-confronting-carbon-inequality-210920-en.pdf.

Paavola, J., and W. N. Adger. 2006. "Fair Adaptation to Climate Change." Ecological Economics 56: 594–609.

Partington, R. 2019. "Bank of England Boss Says Global Finance Is Funding 4C Temperature Rise." The Guardian, October 15, https://www.theguardian.com/business/2019/oct/15/bank-of-england-boss-warns-global-finance-it-is-funding-climate-crisis.

Pasternak, A. 2014. "Voluntary Benefits from Wrongdoing." Journal of Applied Philosophy 31 (4): 377–391.

Paterson, M. 2010. "Legitimation and Accumulation in Climate Change Governance." *New Political Economy* 15 (3): 345–368.

Patterson, J. J., T. Thaler, M. Hoffmann, S. Hughes, A. Oels, E. Chu, and A. Jordan. 2018. "Political Feasibility of 1.5°C Societal Transformations: The Role of Social Justice." *Current Opinion in Environmental Sustainability* 31: 1–9.

Pellizzoni, L. 2004. "Responsibility and Environmental Governance." *Environmental Politics* 13 (3): 541–565.

Pellizzoni, L., and M. Ylönen. 2008. "Responsibility in Uncertain Times: An Institutional Perspective on Precaution." *Global Environmental Politics* 8 (3): 51–7?.

Petroni, M., D. Hill, L. Younes, L. Barkman, S. Howard, I. B. Howell, and M. B. Collins. 2020. "Hazardous Air Pollutant Exposure as a Contributing Factor to COVID-19 Mortality in the United States." *Environmental Research Letters* 15 (9): 0940a9.

Pettit, P. 2007. "Responsibility Incorporated." *Ethics* 117 (2): 171–201.

Pfrommer, T., T. Goeschl, A. Proelss, M. Carrier, J. Lenhard, H. Martin, and H. Schmidt. 2019. "Establishing Causation in Climate Litigation: Admissibility and Reliability." *Climatic Change* 152 (1): 67–84.

Phelan, L., A. Henderson-Sellers, and R. Taplin. 2013. "The Political Economy of Addressing the Climate Crisis in the Earth System: Undermining Perverse Resilience." *New Political Economy* 18 (2): 198–226.

Pierson, B. 2018. "Oil Majors Win Dismissal of New York City Climate Lawsuit." *Reuters*, July 19, https://www.reuters.com/article/us-new-york-climatechange-lawsu t /oil-majors-win-dismissal-of-new-york-city-climate-lawsuit-idUSKBN1K931T.

Pindyck, R. S. 2019. "The Social Cost of Carbon Revisited." *Journal of Environmental Economics and Management* 94: 140–160.

Pogge, T. 2009. *World Poverty and Human Rights: Cosmopolitan Responsibilities and Reforms.* Cambridge, UK: Polity.

Pope Francis. 2015. *Encyclical Letter: Laudato Si' of the Holy Father Francis on Care for Our Common Home.* Rome: Vatican Press. https://www.vatican.va/content/dam/francesco /pdf/encyclicals/documents/papa-francesco_20150524_enciclica-laudato-si_en.pdf.

Pound, J. 2020. "*Cramer Sees Oil Stocks in the 'Death Knell Phase,' Says They Are the New Tobacco.*" CNBC Investing, January 31, https://www.cnbc.com/2020/01 /31/cramer-sees-oil-stocks-in-the-death-knell-phase-says-new-tobacco.html.

Princen, T., and A. Santana. 2015. "Exit Strategies." In *Ending the Fossil Fuel Era*, ed. T. Princen, J. P. Manno, and P. L. Martin, 311–331. Cambridge, MA: MIT Press.

Pullella, P. 2020. *"Vatican Urges Catholics to Drop Investments in Fossil Fuels, Arms."* Reuters, June 18, https://www.reuters.com/article/us-vatican-environment/vatican-urges-catholics-to-drop-investments-in-fossil-fuels-arms-idUSKBN23P1HI.

RAN (Rainforest Action Network) et al. 2020. "Principles for Paris-Aligned Financial Institutions: Climate Impact, Fossil Fuels and Deforestation." https://www.ran.org/wp-content/uploads/2020/09/RAN_Principles_for_Paris-Aligned_Financial_Institutions.pdf.

RAN et al. 2021. *Banking on Climate Chaos: Fossil Fuel Finance Report 2021.* https://www.ran.org/bankingonclimatechaos2021/.

Raval, A., B. Nauman, and G. Tett. 2020. "BP Chief Sees Risk of Oil Demand Passing Peak as Pandemic Hits." *Financial Times*, May 12, https://www.ft.com/content/21affff2-1e57-4000-a439-62cfef6344fb.

Rawlinson, K. 2020. "Lloyd's of London and Greene King to Make Slave Trade Reparations." *The Guardian*, June 17, https://www.theguardian.com/world/2020/jun/18/lloyds-of-london-and-greene-king-to-make-slave-trade-reparations.

Repsol. 2019. "Repsol Will Be a Net Zero Emissions Company by 2050." March 2, https://www.repsol.com/en/press-room/press-releases/2019/repsol-will-be-a-net-zero-emissions-company-by-2050.cshtml.

Reuters. 2020. "Coronavirus Crisis a 'Game Changer' for Oil Sector: Goldman Sachs." *Reuters, March 30*, https://www.reuters.com/article/us-global-oil-research-goldman/coronavirus-crisis-a-game-changer-for-oil-sector-goldman-sachs-idUSKBN21H197.

Revelle, R., and H. E. Suess. 1957. "Carbon Dioxide Exchange between Atmosphere and Ocean and the Question of an Increase of Atmospheric CO_2 during the Past Decades." *Tellus* 9 (1): 18–27.

Rich, N. 2018. "Losing Earth: The Decade We Almost Stopped Climate Change." *New York Times Magazine*, August 1, https://www.nytimes.com/interactive/2018/08/01/magazine/climate-change-losing-earth.html.

Ricke, K., L. Drouet, K. Caldeira, and M. Tavoni. 2018. "Country-Level Social Cost of Carbon." *Nature Climate Change* 8 (10): 895.

Roberts, C., F. W. Geels, M. Lockwood, P. Newell, H. Schmitz, B. Turnheim, and A. Jordan. 2018. "The Politics of Accelerating Low-Carbon Transitions: Towards a New Research Agenda." *Energy Research & Social Science* 44: 304–311.

Roberts, D. 2019a. "Could Squeezing More Oil Out of the Ground Help Fight Climate Change?" *Vox*, December 6, https://www.vox.com/energy-and-environment/2019/10/2/20838646/climate-change-carbon-capture-enhanced-oil-recovery-eor?fbclid=IwAR1zA43fr-AnX8B3L4xsSrXsx7CY581JdeU-kzTKWIFfbt_WaHBF0Geb7nM.

Roberts, D. 2019b. "Pulling CO2 out of the Air and Using It Could Be a Trillion-Dollar Business." *Vox*, November 22, https://www.vox.com/energy-and-environment/2019/9/4/20829431/climate-change-carbon-capture-utilization-sequestration-ccu-ccs.

Robinson, E., and R. C. Robbins. 1968. "*Sources Abundance, and Fate of Gaseous Atmospheric Pollutants.*" US Department of Energy Office of Scientific and Technical Information, https://www.osti.gov/scitech/biblio/6852325.

Robinson, E., and R. C. Robbins. 1969. *Sources, Abundance, and Fate of Gaseous Atmospheric Pollutants*. Final Report and Supplement. https://www.osti.gov/biblio/6852325.

Rogelj, J., O. Geden, A. Cowie, and A. Reisinger. 2021. "Net-Zero Emissions Targets Are Vague: Three Ways to Fix." *Nature* 591: 365–368.

Rogelj, J., G. Luderer, R. C. Pietzcker, E. Kriegler, M. Schaeffer, V. Krey, and K. Riahi. 2015. "Energy System Transformations for Limiting End-of-Century Warming to below 1.5°C." *Nature Climate Change* 5 (6): 519–527.

Rosenbloom, D., H. Berton, and J. Meadowcroft. 2016. "Framing the Sun: A Discursive Approach to Understanding Multi-Dimensional Interactions within Socio-Technical Transitions through the Case of Solar Electricity in Ontario, Canada." *Research Policy* 45 (6): 1275–1290.

Ross, M. L. 2019. "What Do We Know about Export Diversification in Oil-Producing Countries?" *Extractive Industries and Society* 6 (3): 792–806.

Rotberg, R. I., and D. Thompson, eds. 2000. *Truth v. Justice: The Morality of Truth Commissions*. Princeton, NJ: Princeton University Press.

Roth, S. 2019. "California Set a Goal of 100% Clean Energy, and Now Other States May Follow Its Lead." *Los Angeles Times*, January 10, https://www.latimes.com/business/la-fi-100-percent-clean-energy-20190110-story.html.

Rowell, A. 2020. "Is BlackRock Announcement 'Beginning of the End for Fossil-Fuel System?'" *Oil Change International*, January 17, http://priceofoil.org/2020/01/17/is-blackrock-announcement-beginning-of-the-end-for-fossil-fuel-system/.

Rystad Energy. 2020a. "Global E&P Capex Will Reach at Least a 13-Year Low in 2020 as COVID-19 and Price War Persist." March 30, https://www.rystadenergy.com/newsevents/news/press-releases/global-ep-capex-will-reach-at-least-a-13-year-low-in-2020-as-covid-19-and-price-war-persist/.

Rystad Energy. 2020b. "Rystad Energy's Annual Review of World Oil Resources: Recoverable Oil Loses 282 Billion Barrels as Covid-19 Hastens Peak Oil." June 17, https://www.rystadenergy.com/newsevents/news/press-releases/rystad-energys-annual-review-of-world-oil-resources-recoverable-oil-loses-282-billion-barrels-as-covid-19-hastens-peak-oil/.

Sabherwal, A., M. T. Ballew, S. Der Linden, A. Gustafson, M. H. Goldberg, E. W. Maibach, and A. Leiserowitz. 2021. "The Greta Thunberg Effect: Familiarity with Greta Thunberg Predicts Intentions to Engage in Climate Activism in the United States." *Journal of Applied Social Psychology* 51: 321–333.

Sabin Center for Climate Change Law. 2021. "U.S. Climate Change Litigation: Common Law Claims; 28 Cases Found." http://climatecasechart.com/case-category/common-law-claims/.

Sacchi, S., P. Riva, M. Brambilla, and M. Grasso. 2014. "Moral Reasoning and Climate Change Mitigation: The Deontological Reaction toward the Market-Based Approach." *Journal of Environmental Psychology* 38: 252–261.

Sæverud, I. A., and J. B. Skjærseth. 2007. "Oil Companies and Climate Change: Inconsistencies between Strategy Formulation and Implementation?" *Global Environmental Politics* 7 (3): 42–62.

Sanders, B. 2019. *"The Green New Deal."* https://berniesanders.com/en/issues/green-new-deal/.

Santer, B. D., C. J. Bonfils, Q. Fu, J. C. Fyfe, G. C. Hegerl, C. Mears, and C. Z. Zou. 2019. "Celebrating the Anniversary of Three Key Events in Climate Change Science." *Nature Climate Change* 9 (3): 180.

Sanzillo, T. 2021. *"IEEFA: Major Investment Advisors BlackRock and Meketa Provide a Fiduciary Path through the Energy Transition."* March 22, https://ieefa.org/major-investment-advisors-blackrock-and-meketa-provide-a-fiduciary-path-through-the-energy-transition/.

Sardo, M. A. 2020. "Responsibility for Climate Justice: Political Not Moral." *European Journal of Political Theory*. https://doi:10.1177/1474885120955148.

Savage, K. 2019. "Shareholders Sue Exxon for Misrepresenting Climate Risks." The Climate Docket, August 7, https://www.climateliabilitynews.org/2019/08/07/exxon-shareholders-climate-risks-lawsuit/.

Savage, K. 2020. "Maui: Oil Companies Should Pay for Climate Damages They Caused." *The Climate Docket*, October 12, https://www.climatedocket.com/2020/10/12/maui-climate-damages-lawsuit/.

Schumpeter, J. A. 1942. *Capitalism, Socialism and Democracy*. New York: Harper and Brothers.

Scott, A. 2018. "A Managed Decline of Fossil Fuel Production: The Paris Goals Require No New Expansion and a Managed Decline of Fossil Fuel Production." *Oil Change International* Publication Series, *Ecology*, 44 (1). https://www.boell.de/sites/default/files/radical_realism_for_climate_justice_volume_44_1.pdf.

SEI, IISD, ODI, E3G, and UNEP. 2020. *The Production Gap Report*. http://productiongap.org/2020report.

Seto, K. C., S. J. Davis, R. B. Mitchell, E. C. Stokes, G. Unruh, and D. Ürge-Vorsatz. 2016. "Carbon Lock-in: Types, Causes, and Policy Implications." *Annual Review of Environment and Resources* 41: 425–452.

Setzer, J., and C. Higham. 2021. *Global Trends in Climate Change Litigation: 2021 Snapshot. London: Grantham Research Institute on Climate Change and the Environment and Centre for Climate Change Economics and Policy*. London: London School of Economics and Political Science.

Shell. 2020a. "Annual Report and Form 20-F 2018." https://reports.shell.com/annual-report/2018/strategic-report/strategy-business-and-market-overview/risk-factors.php.

Shell. 2020b. "Tackling Climate Change." https://www.shell.com/energy-and-innovation/the-energy-future/shells-ambition-to-be-a-net-zero-emissions-energy-business.html#iframe=L3dlYmFwcHMvY2xpbWF0ZV9hbWJpdGlbi8.

Shell. 2021. *Powering Progress: Annual Report and Accounts for the Year Ended December 31*. https://reports.shell.com/annual-report/2020/servicepages/downloads/files/shell-annual-report-2020.pdf.

Shepherd, T. G. 2016. "A Common Framework for Approaches to Extreme Event Attribution." *Current Climate Change Reports* 2 (1): 28–38.

Sheppard, D. 2020. "Pandemic Crisis Offers Glimpse into Oil Industry's Future." *Financial Times*, May 3, https://www.ft.com/content/99fc40be-83aa-11ea-b872-8db 45d5f6714.

Shue, H. 1996. *Basic Rights: Subsistence, Affluence, and US Foreign Policy.* 2nd ed. Princeton, NJ: Princeton University Press.

Shue, H. 1999. "Global Environment and International Inequality." *International Affairs* 75: 531–545.

Shue, H. 2011. "Human Rights, Climate Change, and the Trillionth Ton." In *The Ethics of Global Climate Change*, ed. D. G. Arnold, 292–314. Cambridge: Cambridge University Press.

Shue, H. 2015. "Historical Responsibility, Harm Prohibition, and Preservation Requirement: Core Practical Convergence on Climate Change." *Moral Philosophy and Politics* 2 (1): 7–31.

Shue, H. 2017. "Responsible for What? Carbon Producer CO_2 Contributions and the Energy Transition." *Climatic Change* 144 (4): 591–596.

Shulman, S. 2012. *Establishing Accountability for Climate Change Damages: Lessons from Tobacco Control.* Union of Concerned Scientists and Climate Accountability Institute, https://www.ucsusa.org/sites/default/files/attach/2016/04/establishing -accountability-climate-change-damages-lessons-tobacco-control.pdf.

Shulman, S. 2017. "Suing Oil Companies to Pay for Climate Change?" *Bulletin of the Atomic Scientists*, October 5, https://thebulletin.org/2017/10/suing-oil -companies-to-pay-for-climate-change/#.

Sinn, H. W. 2008. "Public Policies against Global Warming: A Supply Side Approach." *International Tax and Public Finance* 15 (4): 360–394.

Slater, T., I. R. Lawrence, I. N. Otosaka, A. Shepherd, N. Gourmelen, L. Jakob, and P. Nienow. 2021. "Earth's Ice Imbalance." *The Cryosphere* 15 (1): 233–246.

Small, M., and C. Farand. 2018. *What 30 Years of Documents Show Shell Knew about Climate Science.* DeSmog, May 17, http://www.desmog.co.uk/2018/05/17 /shell-knew-charting-thirty-years-corporate-climate-denialism.

Smee, B. 2019. "Leading Australian Engineers Turn Their Backs on New Fossil Fuel Projects." *The Guardian*, October 20, https://www.theguardian.com/environment

/2019/oct/21/leading-engineers-turn-their-backs-on-new-fossil-fuel-projects
?CMP=share_btn_tw.

Smith, A., A. Stirling, and F. Berkhout. 2005. "The Governance of Sustainable Socio-Technical Transitions." *Research Policy* 34 (10): 1491–1510.

Somanathan, E., T. Sterner, and T. Sugiyama. 2014. "Chapter 15: National and Sub-National Policies and Institutions." In *Climate Change 2014: Mitigation of Climate Change. Contribution of Working Group III to the Fifth Assessment Report of the Intergovernmental Panel on Climate Change*, ed. O. Edenhofer, R. Pichs-Madruga, and Y. Sokona, 1141–1205. Cambridge: Cambridge University Press.

Sovacool, B. K., and M. C. Brisbois. 2019. "Elite Power in Low-Carbon Transitions: A Critical and Interdisciplinary Review." *Energy Research & Social Science* 57: 101242.

Sovacool, B. K., and S. Griffiths. 2020. "Culture and Low-Carbon Energy Transitions." *Nature Sustainability* 3: 685–693.

Sovacool, B. K. 2021. "Who Are the Victims of Low-Carbon Transitions? Towards a Political Ecology of Climate Change Mitigation." *Energy Research & Social Science* 73: 101916.

Spector, M. 2021. "Exxon, Chevron CEOs Discussed Merger in Early 2020—Sources." *Reuters*, January 21, https://www.reuters.com/article/uk-chevron-m-a-exxon-mobil -idUSKBN2A00T3.

Steffen, W., J. Rockström, K. Richardson, T. M. Lenton, C. Folke, D. Liverman, . . . and J. F. Donges, J. 2018. "Trajectories of the Earth System in the Anthropocene." *Proceedings of the National Academy of Sciences* 115 (33): 8252–8259.

Stern, P. C., J. H. Perkins, R. E. Sparks, and R. A. Knox. 2016. "The Challenge of Climate-Change Neoskepticism." *Science* 353 (6300): 653–654.

Stevens, P. 2019. "Exxon Found Not Guilty in New York Climate-Change Securities Fraud Trial, Ending 4-Year Saga." *CBCB*, December 12, https://www.cnbc.com/2019 /12/10/exxon-did-not-mislead-investors-a-new-york-judge-ruled-on-tuesday.html.

Stone, D. A., S. M. Rosier, and D. J. Frame. 2021. "The Question of Life, the Universe and Event Attribution." *Nature Climate Change* 11: 276–278.

Storrow, B. 2019. *"Can a Big Oil Company Go Carbon-Free?"* *Scientific American*, December 6, https://www.scientificamerican.com/article/can-a-big-oil-company -go-carbon-free/.

Stott, M. 2020. "Venezuelan Oil Could Become World's Biggest Stranded Asset, Say Experts." *Financial Times,* November 18, https://www.ft.com/content /cafbd3c7-2434-4f23-8da8-1f7052efdc8e.

Strassel, K. A. 2019. "The Left vs. the Crazy Left." *Wall Street Journal,* August 1, https://www.wsj.com/articles/the-left-vs-the-crazy-left-11564701434.

Stuart-Smith, R. F., F. E. Otto, A. I. Saad, G. Lisi, P. Minnerop, K. C. Lauta, K. V. Zwieten, and T. Wetzer. 2021. "Filling the Evidentiary Gap in Climate Litigation." *Nature Climate Change* 11 (8): 651–655.

Suess, H. E. 1955. "Radiocarbon Concentration in Modern Wood." *Science* 122 (3166): 415–417.

Sullivan, K. 2019. "Could Insurance Lawsuits against Big Oil Be the Next Wave in Climate Liability?" *The Climate Docket,* July 18, https://www.climatedocket.com /2019/07/18/insurance-lawsuits-climate-liability/.

Sullivan, R., and A. Gouldson. 2017. "The Governance of Corporate Responses to Climate Change: An International Comparison." *Business Strategy and the Environment* 26 (4): 413–425.

Sun, Q., et al. 2021. "A Global, Continental, and Regional Analysis of Changes in Extreme Precipitation." *Journal of Climate* 34 (1): 243–258.

Sunstein, C. R. 1997. *Free Markets and Social Justice.* New York: Oxford University Press.

Supran, G., and N. Oreskes. 2017. "Assessing ExxonMobil's Climate Change Communications (1977–2014)." *Environmental Research Letters* 12 (8): 084019.

Supran, G., and N. Oreskes. 2021. "Rhetoric and Frame Analysis of ExxonMobil's Climate Change Communications." *One Earth* 4 (5): 696–719.

Swyngedouw, E. 2010. "Apocalypse Forever? Post-Political Populism and the Spectre of Climate Change." *Theory, Culture & Society* 27 (2–3): 213–232.

Tabuchi, H. 2020. "How One Firm Drove Influence Campaigns Nationwide for Big Oil." *New York Times,* November 11, https://www.nytimes.com/2020/11/11 /climate/fti-consulting.html.

Taylor, M., and J. Ambrose. 2020. "Revealed: Big Oil's Profits since 1990 Total Nearly \$2tn." *The Guardian,* February 12, https://www.theguardian.com/business /2020/feb/12/revealed-big-oil-profits-since-1990-total-nearly-2tn-bp-shell chevron-exxon.

Thomas, J. A. 2017. "Confronting Climate Change: The Uneasy Alliance of Scientists and Nonscientists in a Neoliberal World." *Environmental History* 23 (1): 172–182.

Tillmann, T., J. Currie, A. Wardrope, and D. McCoy. 2015. "Fossil Fuel Companies and Climate Change: The Case for Divestment." *BMJ* 350. https://www.bmj.com/content/350/bmj.h3196.

Tol, R. S. J. 2019. "A Social Cost of Carbon for (Almost) Every Country." *Energy Economics* 83: 555–566.

Tooze, A. 2019. "Why Central Banks Need to Step Up on Global Warming. A Decade after the World Bailed Out Finance, It's Time for Finance to Bail out the World." *Foreign Policy*, July 20, https://foreignpolicy.com/2019/07/20/why-central-banks-need-to-step-up-on-global-warming/.

TotalEnergies. 2020. "Total Energy Outlook 2020." https://www.total.com/sites/g/files/nytnzq111/files/documents/2020-09/total-energy-outlook-presentation-29-september-2020.pdf.

Transparency International. 2020. "Corruption Perceptions Index." https://www.transparency.org/en/cpi/2020/index/nzl.

Trenberth, K. E., L. Cheng, P. Jacobs, Y. Zhang, and J. Fasullo. 2018. "Hurricane Harvey Links to Ocean Heat Content and Climate Change Adaptation." *Earth's Future* 6 (5): 730–744.

Turnheim, B., and F. W. Geels. 2012. "Regime Destabilisation as the Flipside of Energy Transitions: Lessons from the History of the British Coal Industry (1913–1997)." *Energy Policy* 50: 35–49.

Turnheim, B., and F. W. Geels. 2013. "The Destabilisation of Existing Regimes Confronting a Multi-Dimensional Framework with a Case Study of the British Coal Industry (1913–1967)." *Research Policy* 42 (10): 1749–1767.

UCS (Union of Concerned Scientists). 2016. *Ending ExxonMobil Sponsorship of the American Geophysical Union.* Cambridge, MA: Union of Concerned Scientists.

UCS. 2018a. "The Disinformation Playbook." https://www.ucsusa.org/our-work/center-science-and-democracy/disinformation-playbook#.WzJwuql9ijQ.

UCS. 2018b. *The Science Connecting Extreme Weather to Climate Change.* Cambridge, MA: Union of Concerned Scientists. https://www.ucsusa.org/our-work/global-warming/science-and-impacts/climate-attribution-science.

UN (United Nations). 2018. *Framework Principles of Human Rights and the Environment.* New York: Human Rights Office of the High Commissioner.

UNCTAD (United Nations Conference on Trade and Development). 2019. *Trade and Development Report 2019.* Geneva: UNCTAD.

UNDDR (United Nations Office for Disaster Risk Reduction) and CRED (Centre for Research on the Epidemiology of Disasters). 2020. *The Human Cost of Disasters: An Overview of the Last 20 Years (2000–2019).* Geneva: UNDDR.

UNFCCC. 1992. *United Nations Framework Convention on Climate Change.* https://unfccc.int/resource/docs/convkp/conveng.pdf.

UNFCCC. 2016. *Just Transition of the Workforce, and the Creation of Decent Work and Quality Jobs.* United Nations technical paper, https://unfccc.int/sites/default/files/resource/Just%20transition_for%20posting.pdf.

University of California San Francisco Library. 1969. Author unknown, document late 1969 August 21, letter/memo, ID xqkd0134, box 555, file NFI, case 100042026. Brown & Williamson Records Collection, https://www.industrydocumentslibrary.ucsf.edu/tobacco/docs/#id=xqkd0134.

Unruh, G. C. 2000. "Understanding Carbon Lock-In." *Energy Policy* 28 (12): 817–830.

Unruh, G. C. 2002. "Escaping Carbon Lock-In." *Energy Policy* 30 (4): 317–325.

Urban, M. C. 2015. "Accelerating Extinction Risk from Climate Change." *Science* 348 (6234): 571–573.

US Department of State. 2021. "Fireside Chat at IHS CERAWeek." March 2, https://www.state.gov/fireside-chat-at-ihs-ceraweek/.

Vanderheiden, S. 2011. "Globalizing Responsibility for Climate Change." *Ethics & International Affairs* 25 (1): 65–84.

Vanderheiden, S. 2016. "Climate Justice beyond International Burden Sharing." *Midwest Studies In Philosophy* 40 (1): 27–42.

Vardi, I. 2018. *"Fossil Fuel Industry Outspent Environmentalists and Renewables by 10:1 on Climate Lobbying, New Study Finds."* DeSmog, July 18, https://www.desmog.co.uk/2018/07/18/fossil-fuel-industry-outspent-environmentalists-renewables-10-1-climate-lobbying-study.

Vautard, R., O. Boucher, G. J. van Oldenborgh, F. Otto, K. Haustein, M. M. Vogel, S. I. Seneviratne, J.-M. Soubeyroux, M. Schneider, A. Drouin, A. Ribes, F. Kreienkamp,

P. Stott, and M. van Aalst. 2019. *"Human Contribution to the Record-Breaking July 2019 Heat Wave in Western Europe."* World Weather Attribution, August 2, https://www.worldweatherattribution.org/human-contribution-to-the-record-breaking-july-2019-heat-wave-in-western-europe/.

Verisk Maplecroft. 2021. *Political Risk Outlook 2021.* March 25, https://www.maplecroft.com/insights/analysis/political-risk-outlook-2021/.

Victor, D. G., D. H. Hults, and M. C. Thurber. 2012a. *"Introduction and Overview."* In *Oil and Governance: Oil and Governance: State-Owned Enterprises and the World Energy Supply,* ed. D. G. Victor, D. H. Hults, and M. C. Thurber, 3–31. Cambridge: Cambridge University Press.

Victor, D. G., D. H. Hults, and M. C. Thurber. 2012b. *"Major Conclusions and Implications for the Future of the Oil Industry."* In *Oil and Governance: State-Owned Enterprises and the World Energy Supply,* ed. D. G. Victor, D. H. Hults, and M. C. Thurber, 887–928. Cambridge: Cambridge University Press.

Vohra, K., A. Vodonos, J. Schwartz, E. A. Marais, M. P. Sulprizio, and L. J. Mickley. 2021. "Global Mortality from Outdoor Fine Particle Pollution Generated by Fossil Fuel Combustion: Results from GEOS-Chem." *Environmental Research* 195: 110754.

Volcovici, V. 2018. "Big Oil Eyes U.S. Minority Groups to Build Offshore Drilling Support." *Reuters,* June 22, https://www.reuters.com/article/us-usa-oil-offshore/big-oil-eyes-u-s-minority-groups-to-build-offshore-drilling-support-idUSKBN1JI2FM.

Wagner, D. L., C. Kuveke, L. Ross, and A. Zibel. 2020. "Bailed Out & Propped Up: U.S. Fossil Fuel Pandemic Bailouts Climb Toward $15 Billion." Bailout Watch, *Public Citizen, and Friends of the Earth,* November, https://report.bailoutwatch.org/.

Wagner, G., D. Anthoff, M. Cropper, S. Dietz, K. T. Gillingham, B. Groom, and J. H. Stock. 2021. "Eight Priorities for Calculating the Social Cost of Carbon." *Nature* 590: 548–550.

Wallimann-Helmer, I., L. Meyer, K. Mintz-Woo, T. Schinko, and O. Serdeczny. 2019. "Ethical Challenges in the Context of Climate Loss and Damage." In *Loss and Damage from Climate Change. Concepts, Methods and Policy Options,* ed. R. Mechler, L. Bouwer, T. Schinko, and S. Surminski, 39–61. Dordrecht: Springer.

Warlenius, R. 2018. "Decolonizing the Atmosphere: The Climate Justice Movement on Climate Debt." *Journal of Environment & Development* 27 (2): 131–155.

Watts, J., J. Ambrose, and A. Vaughan. 2019. "Oil Firms to Pour Extra 7m Barrels per Day into Markets, Data Shows." *The Guardian,* October 10, https://www.theguardian.com/environment/2019/oct/10/oil-firms-barrels-markets.

Watts, M. J. 2005. "Righteous Oil? Human Rights, the Oil Complex, and Corporate Social Responsibility." *Annual Review of Environment and Resources* 30: 373–407.

Watts, N., M. Amann, N. Arnell, S. Ayeb-Karlsson, J. Beagley, K. Belesova, and A. Costello. 2021. "The 2020 Report of The Lancet Countdown on Health and Climate Change: Responding to Converging Crises." *The Lancet* 397 (10269): 129–170.

Welsby, D., J. Price, S. Pye, and P. Ekins. 2021. "Unextractable Fossil Fuels in a 1.5 °C World." *Nature* 597, 230–234.

Wenar, L. 2011. "Clean Trade in Natural Resources." *Ethics & International Affairs* 25 (1): 27–39.

Wikipedia. n.d.a. Lukoil. https:// https://en.wikipedia.org/wiki/Lukoil. Accessed on September 25, 2021.

Wikipedia. n.d.b. National Iranian Oil Company. https://en.wikipedia.org/wiki /National_Iranian_Oil_Company. Accessed on September 25, 2021.

Williams, E. 2020. "Attributing Blame?—Climate Accountability and the Uneven Landscape of Impacts, Emissions, and Finances." *Climatic Change* 161: 273–290.

WMO (World Meteorological Organization). 2021. *WMO Statement on the State of the Global Climate 2020.* Geneva: WMO.

World Bank. 2019. "Oil Rents (% of GDP)." https://data.worldbank.org/indicator /NY.GDP.PETR.RT.ZS.

WRI (World Resource Institute). n.d. "Greenhouse Gas Protocol." https:// ghgprotocol.org/. Accessed on September 25, 2021.

Wright, C., and D. Nyberg. 2014. "Creative Self-Destruction: Corporate Responses to Climate Change as Political Myths." *Environmental Politics* 23 (2): 205–223.

XR. n.d. Extinction Rebellion. https://rebellion.global/about-us/. Accessed on September 25, 2021.

Young, I. M. 2006. "Responsibility and Global Justice: A Social Connection Model." *Social Philosophy and Policy* 23 (1): 102–130.

Za, V. 2019. *Italy's UniCredit to Exit Thermal Coal Financing by 2023.* Reuters, November 26, https://www.reuters.com/article/us-unicredit-coal-exit-idUSKBN1Y01UA.

INDEX